DATA VISUALIZATION FOR SCIENTIFIC DISCOVERIES

Technological Integration and Innovative Practices

DATA VISUALIZATION FOR SCIENTIFIC DISCOVERIES

Technological Integration and Innovative Practices

Qian Liu
Xi'an Jiaotong-Liverpool University, China

Wenling Li
Xi'an Eurasia University, China

World Scientific

NEW JERSEY • LONDON • SINGAPORE • BEIJING • SHANGHAI • HONG KONG • TAIPEI • CHENNAI • TOKYO

Published by

World Scientific Publishing Co. Pte. Ltd.
5 Toh Tuck Link, Singapore 596224
USA office: 27 Warren Street, Suite 401-402, Hackensack, NJ 07601
UK office: 57 Shelton Street, Covent Garden, London WC2H 9HE

Library of Congress Cataloging-in-Publication Data
Names: Liu, Qian (Of Xi'an Jiaotong-Liverpool University) author | Li, Wenling, 1963– author
Title: Data visualization for scientific discoveries : technological integration and innovative practices / Qian Liu, Xi'an Jiaotong-Liverpool University, China, Wenling Li, Xi'an Eurasia University, China.
Description: New Jersey : World Scientific, [2027] | Includes bibliographical references.
Identifiers: LCCN 2026001721 | ISBN 9789819827794 hardcover |
ISBN 9789819827800 ebook for institutions | ISBN 9789819827817 ebook for individuals
Subjects: LCSH: Communication in science | Information visualization
Classification: LCC Q223 .L574 2026
LC record available at https://lccn.loc.gov/2026001721

British Library Cataloguing-in-Publication Data
A catalogue record for this book is available from the British Library.

For any available supplementary material, please visit
https://www.worldscientific.com/worldscibooks/10.1142/14719#t=suppl

Desk Editors: Eshak Nabi Akbar Ali/Veronica Lee

Typeset by Stallion Press
Email: enquiries@stallionpress.com

Preface I — Data Visualization and Scientific Discovery: The Cognitive Revolution in the Age of Digital Media

In this era of data explosion, we are experiencing an unprecedented cognitive revolution. As a long-time research and teaching practitioner in the intersection of digital media and science and art, I am deeply aware that data visualization is evolving from a simple tool to a new form of scientific language and artistic expression. This book was born out of a long-term accumulation of observations and practices on this emerging field of possibilities.

Traditional data visualization is often limited to static graphs and simple 2D presentations, and the complexity and diversity of modern scientific research pose a serious challenge to this traditional approach. Take climate models as an example, a global climate simulation may contain tens of millions of grid points, each of which contains multiple variables such as temperature, humidity, and wind speed. These data not only are huge in scale but also have obvious characteristics of space-time dynamics. This is particularly true of protein folding studies, which look at protein conformations at the atomic scale, with each frame containing precise location information for hundreds of thousands of atoms. In the face of such data and Lex scientific data, traditional visualization methods have proved inadequate.

The rapid development of artificial intelligence technology has brought revolutionary breakthroughs in data visualization. Taking AlphaFold as an example, the opal structure prediction system developed by DeepMind not only accurately predicts the three-dimensional structure of opal but also dynamically visualizes the entire process of protein folding. This technology, which combines deep learning with 3D rendering, allows scientists to visually observe and understand protein conformations, providing a whole new perspective on drug design and disease research.

The interactive nature of digital media injects new energy into scientific visualization. Virtual reality technology allows researchers to "go inside" molecules to observe the interaction of atoms from an arbitrary angle; augmented reality technology superimposes scientific data on a real-world scene to create a unique mixed reality experience. In teaching, we use VR technology to let students "experience" the meteorite impact when the Acrophobia died out. This immersive learning experience greatly improves the teaching effect.

The theoretical framework of this book is based on the "four-dimensional model of scientific visualization." This model systematically expounds the core elements of modern scientific visualization from four aspects: data dimension, visual dimension, interaction dimension, and narrative dimension. In the data dimension, we are concerned with how to accurately express the spatial and temporal characteristics and multivariate relationships of scientific data. In the visual dimension, we explore the scientific application of visual elements such as color coding and graphic metaphor. In the interactive dimension, we study the mechanisms of dynamic dialogue between the user and the visualization system. In the narrative dimension, we explore how to tell compelling science stories through visualization.

In terms of technical practice, this book provides a detailed introduction to current state-of-the-art visualization tools and methods. From Web-based dynamic visualization libraries D3.js and Three.js, to professional scientific computing visualization tool ParaView; from the game engine Unity 3D in scientific visualization, to the latest advances in artificial intelligence. We focus on how these techniques can be applied to specific problems in science.

Transboundary integration is an important theme in this book. We present many examples of innovation from research, engineering, and the arts. In science, protein-folding games such as Foldit transform a complex problem in science into a game challenge that Volkswagen can take part in. In terms of engineering applications, digital twin technologies are changing models of urban management and industrial design. In the creation of art, scientists and artists collaborate to transform brainwave data into a moving musical visualization. These examples vividly demonstrate how scientific visualization can break down disciplinary boundaries and create new values.

This book is for all readers who are interested in scientific visualization. Whether you're a researcher looking to better present your research, a designer looking for new creative inspiration, or an educator looking for innovative teaching methods, you'll find valuable content in this book. For policymakers, understanding the power of scientific visualization can help them translate complex scientific findings into policy consensus.

As artificial intelligence reshapes the way humans perceive, scientific visualization is at an important turning point. It is no longer merely an aid to scientific research but is becoming a stand-alone way of cognition and an art form. This book hopes to contribute to this ongoing transformation and inspire more innovation and exploration across disciplines. Because the most exciting scientific discovery often begins with a stunning visual moment — the one that makes you feel like you can't help it. "Aha," it was a fantastic moment.

Qian Liu
October 2025

Preface II — Visualization of Data and Scientific Discovery: An Interdisciplinary Cognitive Symphony

Dear reader friends,

When you open this book, we may be in a similar situation — in the face of increasingly complex problems in science, the traditional single-subject perspective has become inadequate. As a long-time scholar of cognitive psychology and interdisciplinary research, I hope to share with you some of my thoughts and explorations at the intersection of data visualization and scientific discovery with this book.

One afternoon last spring, I remember an interesting scene in my office: a postdoctoral fellow in cognitive psychology and a graduate student in computer science were fighting over how to present a set of eye movement and electroencephalography data. Psychologists wanted to turn eye movement and EEG data into graphs that people could understand, and after long explanations and discussions, they finally understood each other's desires and tactics. In the end, they not only produced a visual image that was very satisfying to both parties but also unexpectedly led to an interdisciplinary collaborative research project.

Stories like this have been common throughout my research. Data visualization is like a great interpreter, able to translate the "dialects" of different disciplines into the "Mandarin" that everyone knows. In this book, I want to talk to you about how to build cognitive bridges between different disciplines through visual language.

Cognitive psychology tells us that the way the human brain processes information is surprisingly common. When we see a graceful curve, mathematicians think of functional equations, musicians think of the melody line, and a painter may see a moving outline. This cross-modal cognitive nature is the underlying reason why data visualization can break down disciplinary barriers. In my research, I am particularly interested in using this commonality to create visualizations that maintain scientific rigor while also being artistic.

Those exciting moments of discovery in the lab often stem from interdisciplinary collisions. For example, when we modeled urban traffic flow to simulate neural signal transmission, both engineers and neuroscientists were given new insights. When we turn the graph of a network into an interactive music visualizer, biologists and composers see an unprecedented connection. These examples vividly demonstrate that good visualization not only presents the known but also inspires the unknown.

In scientific collaboration, visualization often resolves professional barriers. I worked on a research project on reading grading that brought together psychologists, computer scientists, and sociologists. Initially, the data and analytical methods varied greatly between groups until we developed a visualization system that integrated reading data, book classification indicators, and social survey results. The system not only led to high-quality research results but also allowed team members to build a deep interdisciplinary understanding.

The process of writing this book is also a compendium of my own academic career. From early years focusing on cognitive psychological experiments, and later expanded to scientific visualization, as a result of my work on interdisciplinary educational innovation, I am increasingly aware that the boundaries of disciplines are often artificially defined, and real-world problems are never stuck within a discipline. Data visualization gives us a unique tool that allows us to look beyond these human boundaries and see the existing connections between knowledge.

In the digital age, we face an unprecedented information overload. How to make huge amounts of data yield insights, how to make complex knowledge easier to understand, how to enable experts in different fields to collaborate effectively — all point to the same answer: we need better visualization methods, smarter visual expressions. This is not only a technological challenge but also a cognitive revolution.

Hopefully, this book will provide you with some inspiration and tools to experience the surprises and benefits of interdisciplinary visualization in your own research or teaching. After all, the best scientific discoveries often come not from the deep wells of one discipline but from the intersection of different fields of knowledge — just as sunlight refracts a brilliant spectrum of colors through prisms.

We look forward to continuing this journey of discovery across disciplinary boundaries with you.

Wenling Li
Fall, 2025

About the Authors

Qian Liu is the Dean of the Academy of Film and Creative Technology (AFCT) and a Professor at Xi'an Jiaotong-Liverpool University (XJTLU), following his tenure as the founding Dean of the Digital Media School at the Beijing Film Academy (BFA). He has held a range of academic research positions in the US and China: visualization integration researcher at the California Institute for Telecommunication and Information Technology (Calit2), University of California, San Diego, as well as system researcher at the National Center for Supercomputing Applications (NCSA), US, and China Electric Power Research Institute (CEPRI).

His diverse research and creative interests span across digital media arts and technology, digital filmmaking, immersive display systems, interactive design, VR/AR/XR, and generative AI applications — an expertise he also practices as an internationally exhibited digital media artist.

Professor Liu has achieved significant, tangible results by applying his creativity and research across his interdisciplinary specializations. He was the former Executive Chairman of the Digital Imaging Creative Committee of the China Video Industry Association (CVIA) and Deputy Director of the 5G Professional Committee of China Communications Industry Association. He also held the position of Chairman of the International Creative Art Contest (ICAC) at the 2024 Busan International Art Festival (BIAF) and Chair of the International AI Visual art Contest

(IAIV) at the 2025 BIAF. Currently, he is a member of the 2024 film gold digital Honor Recommendation Committee, as well as a member of the AI Imaging/Video Talent Selection Programme Recommendation Committee.

Wenling Li is the Chief Expert at the College of Humanities and Education and a Professor at Xi'an Eurasia University, China. She is a distinguished educational researcher with extensive experience in the field, having previously held dual roles as the Director of the Doctoral Program and Doctoral Advisor at U.S.-based Trident International University's College of Education.

Professor Li holds a Ph.D. from Beijing Normal University (1991) and completed her postdoctoral research at the University of Illinois at Urbana-Champaign (1992), where she later served as a Senior Research Scientist at the university's Children's Reading Research Center. During this time, she led the Spencer Foundation-funded comparative reading research project between China and the U.S.

She is currently an active member of several prestigious professional organizations, including the American Educational Research Association (AERA), the National Association for Early Childhood Education (NAEYC), and the American Council on Foreign Language Teaching (ACTFL). Her major publications include one book published in the U.S. and ten books published in China, reflecting her deep scholarly contributions to both countries.

Contents

Part 1

Foundations: Understanding the Symbiosis Between Scientific Visualization and Digital Media

Chapter 1

The Nature of Scientific Discovery and Visualization Needs

1.1. Introduction: The Nature of Scientific Discovery — The Cognitive Transition from "Observation" to "Insight"

The essence of scientific discovery is the process by which human beings gradually reveal the laws of nature and expand the boundaries of cognitive knowledge through systematic observation, experiment, and reasoning. It began with curiosity about phenomena (e.g., Newton's thought about the landing of an apple and Galileo's observation of the moon of Jupiter in a telescope). It is based on the deep excavation of data (such as Mendeleev's induction of the periodic law of the elements and Watson and Crick's analysis of the double helix of DNA) and ultimately leads to the refinement of the universal theory (such as Einstein's theory of relativity and Darwin's theory of evolution) (Tolle, *et al.*, 2011). This process is essentially a cognitive transition from chaos to order — scientific data (whether the values of experimental measurements, observations recorded phenomena, or parameters generated by simulations are like scattered jigsaw puzzles, and the goal of scientific discovery is to piece these jigsaw puzzle pieces into a complete picture, revealing deep laws hidden beneath the surface (Kitchin, 2022).

However, as the complexity of scientific research grows exponentially (e.g., interdisciplinary research becomes the norm and the accuracy of laboratory facilities increases in geometric progression), the “fragmentation” and “sea quantization” characteristics of scientific data are becoming more pronounced: A large particle physics experiment may produce petabytes of collision event data, a high-resolution video microscope can generate terabytes of cellular images per day, and a simulation of climate models requires processing time series data of tens of millions of grid points. These data are no longer simply “observational records,” but rather “high-dimensional information complexes” that contain multi-dimensional variables (such as time, space, and physical parameters), complex associations (such as nonlinear coupling between variables), and dynamic processes (such as the evolutionary sequence of a system). With such large data volumes and complex structural features, the traditional “artificial observation + empirical analysis” model has become difficult to cope with — researchers not only need to quickly sift through huge amounts of data to sift through key information, but there is a need to understand the physical mechanisms and scientific logic behind the data in an intuitive way (Sacha *et al.*, 2020).

This is the core value of scientific visualization: It is a kind of technology and methodology to transform abstract scientific data into intuitive visual expression. It aims to transform the “hidden knowledge” of data into “explicit cognition” through visual languages, such as graphics, images, animation, and so on, thereby accelerating the process of scientific discovery (Ware, 2020). In other words, scientific visualization is not only a “display tool” for data but also a “cognitive magnifying glass” for scientists. It facilitates researchers in overcoming cognitive limitations by providing multi-dimensional visual encoding (such as color mapping to represent variable sizes, streamlines to show flow directions, and three-dimensional models to depict spatial structures), dynamic temporal presentations (like simulating the evolution of a system), and interactive exploration features (such as allowing users to adjust parameters and observe changes in the results). This enables scientists to accurately capture key patterns from the “data ocean” and ultimately bridge the gap from “observing data” to “discovering the essence.” It helps researchers break through cognitive limitations, accurately capture key patterns from the “ocean of data,” and finally achieve the leap from “observing data” to “insight into the nature (Liu & Chen, 2021; Johnson & Roussou, 2022).”

1.2. Paradigm Shifts in Scientific Discovery: From Experience-Driven to Data-Intelligence

The nature of scientific discovery is the process of human beings revealing natural laws and expanding cognitive boundaries through systematic exploration. However, the “methodology” of this process is not immutable — with the progress of technical tools and the explosive growth of data scale, scientific discoveries have undergone a paradigm leap from experience-driven to theoretical deduction to data-intensive exploration. Understanding this evolutionary vein is the key premise of understanding why scientific visualization has become an indispensable cognitive tool in contemporary science (Tolle, *et al.*, 2011).

1.2.1. *The First Paradigm: Empirical Observation and Experimental Science (Ancient–17th Century)*

The earliest scientific habits of human beings stem from the direct observation of natural phenomena and the summary of experience. From the ancient Greek philosophers who proposed the “geocentric theory” by observing the movement of celestial bodies with the naked eye, to the induction of twenty-four solar terms through agricultural practice in ancient China, to Galileo’s observation of Jupiter’s satellites with telescopes, and Kepler’s summary of the three laws of planetary motion based on Tycho’s planetary position data — this stage is called “empirical paradigm” or “Experimental Science Paradigm (Shapin, 2021).”

Its core feature is that scientific knowledge comes from direct observation of natural phenomena, repeated experiments, and inductive summaries. Scientists design controllable experiments (such as Galileo’s slope experiment), record key variables (such as temperature, time, and displacement), and extract causal relationships (such as constant acceleration of free fall) from them. At this time, the amount of data is limited (usually mainly manually recorded tables or text descriptions), and scientific discoveries rely on researchers’ intuition, logical reasoning, and repeated verification.

Limitations: Limited by the accuracy of observation tools (such as naked-eye resolution and error of mechanical instruments) and the repeatability of experiments, many microscopic (such as atomic structure) or macroscopic (such as cosmic evolution) phenomena cannot be directly captured, and the scope of scientific exploration is constrained in the range of "visible, touchable and measurable."

1.2.2. *Second Paradigm: Theoretical Model and Mathematical Derivation (17th–20th century)*

With the improvement of mathematical tools (such as Newton's calculus and Maxwell's equations) and the promotion of philosophical trends (such as Descartes' rationalism), scientists began to try to explain observed phenomena with theoretical models and derive unknown laws through mathematical formulas. This stage is called "theoretical paradigm (Helge, 2021)."

Its core feature is that scientific knowledge is based on abstract theoretical assumptions, and universality conclusions are derived through mathematical logic. For example, Newton proposed the law of universal gravitation $\left(F = G\frac{m_1 m_2}{r^2}\right)$ based on the "absolute view of space-time." Einstein reconstructed the nature of space-time and gravity through relativity, and quantum mechanics used quantum state and probability amplitude to describe the behavior of microscopic particles. These theories can not only explain the existing experimental data but also predict phenomena that have not yet been observed (e.g., the discovery of Neptune comes from the prediction of the orbital deviation of Uranus by Newtonian mechanics).

Data role: At this time, the data are mainly used to verify the correctness of the theory (such as measuring the gravitational constant G through the Cavendish torsional scale experiment), but the construction of the theory itself relies more on the "thought experiments" and logical deductions of mathematicians and physicists. Although the amount of data has increased (such as astronomical observation records and laboratory measurement tables), it is still dominated by manually sorted small-scale datasets, and the visualization form is mostly static charts (such as function curves and geometric diagrams) (Tversky & Morrison, 2021).

1.2.3. *The Third Paradigm: Computational Simulation and Digital Experiments (early 21st century)*

In the middle of the 20th century, the birth of electronic computers completely changed the pattern of scientific research. When some scientific problems (such as nuclear explosion simulation, atmospheric circulation prediction, and protein folding dynamics) cannot be solved directly through traditional experiments or theoretical inference due to the high cost of experiments (such as the inability to detonate a nuclear bomb in real life), too large scale (such as cosmic evolution), and too long (such as biological evolution), scientists begin to use computers which conduct numerical simulation and digital experiments — by establishing mathematical models (such as partial differential equations and statistical mechanics formulas), natural processes are transformed into computer-calculable algorithms, and the behavior of complex systems is restored through large-scale numerical calculations. This stage is called "computational paradigm" or "computer simulation paradigm (Berman *et al.*, 2022; Lynch, 2020)."

Its core feature is that scientific discovery relies on the numerical solution of theoretical models by computers, and some real experiments are replaced by simulation experiments. For example, meteorologists predict typhoon paths by coupling atmospheric-ocean models, materials scientists simulate the electronic structure of high-temperature superconducting materials through molecular dynamics, and bioinformaticians search for disease-related genes through genome comparison algorithms. At this time, the amount of data has increased significantly (a single simulation may produce GB-level or even TB-level output), but the data still exist in the form of "simulation results" and require professional scientists to interpret its physical significance.

The need for visualization has begun to appear: In order to understand the physical meaning of the simulation results (such as the vortex structure in fluid mechanics and the atomic motion trajectory in molecular dynamics), scientists began to use simple two-dimensional charts (such as velocity field cloud diagrams and line diagrams of energy changes over time) or three-dimensional geometric models (such as the sticks of protein molecules). However, these visualizations are mostly "post-analysis tools," which are mainly used to verify the correctness of simulation, rather than actively guiding scientific exploration (Sacha *et al.*, 2020).

1.2.4. *Fourth Paradigm: Data-Intensive Scientific and Intelligent Discovery (21st Century to the Present)*

Entering the 21st century, with the popularization of sensor networks (such as satellite remote sensing and Internet of Things devices), high-throughput experimental technologies (such as gene sequencers generating GB-level data per second and particle colliders recording millions of collisions per second), and Internet digitalization (such as scientific literature databases and social media behavior records), scientific research has entered the "data-intensive paradigm" (also known as the fourth paradigm) — the core driving force of scientific discovery is no longer pre-designed experiments or theoretical assumptions but the mining and analysis of massive, multi-source, and heterogeneous data (Tolle, *et al.*, 2011).

Its core feature is that data themself become the "raw materials" of scientific research. Scientists automatically discover patterns, associations, and assumptions from data through data mining, machine learning, and other technologies and then derive new scientific laws (Chen *et al.*, 2021). For example:

- **Astronomy:** LSST (Large Comprehensive Sky Telescope) will generate 20TB of starry sky image data every night. By analyzing the position and brightness changes of hundreds of millions of stars, scientists have discovered the transit signals of exoplanets or the gravitational lens effect of dark matter.
- **Life science:** The Human Genome Project produced sequence data of 3 billion base pairs. The subsequent single-cell RNA sequencing technology allows a single experiment to produce gene expression spectra of millions of cells, from which researchers need to identify disease-related cell subgroups and key pathways.
- **High-energy physics:** The Large Hadron Collider (LHC) of the European Center for Nuclear Research (CERN) generates more than

30PB of particle collision data every year. Scientists use machine learning algorithms to screen out rare signals of Higgs boson decay from a large number of events (Wilkinson, *et al.*, 2016).

Challenges and opportunities: The data scale (PB level or even EB level), dimension (hundreds to thousands of variables), and dynamism (real-time stream data) under the fourth paradigm far exceed human direct cognitive ability — scientists cannot find laws by scanning the original data with the naked eye, and it is difficult to process high-dimensional correlations with traditional statistical methods. At this time, data visualization has become a key bridge between "raw data" and "scientific insight": It can not only transform complex data into perceptible visual forms (such as three-dimensional galaxy distribution and dynamic gene regulation networks) but also support scientists' active exploration through interactive design (such as scaling, filtering, and association highlighting), unknown patterns in the data, and even generate new research assumptions (Kitchin, 2022).

1.2.5. *The Essence of Paradigm Evolution: From "Data Serves Cognition" To "Cognition Depends on Data"*

Looking back on these four paradigms, the logic of scientific discovery has undergone a fundamental change:

- In the first three paradigms, data are the "results" or "verification tools" of scientific research — experimental data are used to support theory, simulation data are used to supplement experiments, and theory itself is the starting point of cognition.
- In the fourth paradigm, data become the "starting point" and "core driving force" of scientific research — scientists first face a large amount of raw data (which may come from sensors, experimental equipment, or public databases) and then explore the patterns through data cleaning, analysis, and visualization, and the outcome may lead to the development of new theories or the refinement of existing knowledge.

This transformation directly gives rise to the rigid need for "scientific visualization": When the data scale exceeds the limit of human working memory (psychological research shows that humans can only process 4±1 information units in a short period of time), when the association between variables is hidden in high-dimensional space (such as the non-linearity of thousands of genes and disease phenotypes in gene expression data sexual relationship), and when dynamic processes need to be tracked in real time (such as temperature/humidity/ocean current coupling evolution in climate change models), traditional tables, formulas, or static charts can no longer meet the needs — scientists need a tool that can "translate" data into visual language, through color mapping differences and geometry express structure, dynamic interaction to explore associations, and transform abstract numbers into perceptible, operable, and inferable cognitive objects.

The paradigm evolution of scientific discovery is essentially the collaborative evolution of human cognitive tools and the data environment. From empirical science that relies on sensory observation, to theoretical science based on mathematical models, to digital science through computer simulation, to today's intelligent science with data mining as the core, every paradigm leap is accompanied by an exponential growth in data scale and cognitive complexity. And data visualization is the indispensable "cognitive translator" for scientific discovery under the fourth paradigm — it allows scientists to "see" laws in the ocean of data and "touch" the truth in complex systems (Borgman, 2019; Borgman, 2021; Ware, 2020).

1.3. Complexity of Scientific Data: Multi-Source Heterogeneity, High-Dimensional Dynamics, and Uncertainty Challenges

In today's scientific research process, data have become core elements that drive scientific discovery. However, these data are not uniform and easy to process but display a high degree of complexity. This complexity is reflected in the key aspects of multi-source heterogeneity, high-dimensional dynamics, and uncertainty, which pose enormous challenges for scientists to understand and analyze data, while also highlighting the important role of data visualization (Gershon & Page, 2020).

1.3.1. *Multi-Source Heterogeneity: A Multifaceted Puzzle of Scientific Data*

Modern scientific research involves many different fields and directions, and its data sources are extremely wide, covering sensors, analog computing, text data, and other forms. These data have significant differences in type and structure, forming multi-source heterogeneous features (Ma & Liu, 2021).

Sensors and experimental instruments are important sources of scientific data. In the study of many disciplines such as physics, chemistry, and biology, various sensors and experimental instruments are used to directly measure and record various physical quantities in natural phenomena or in the experimental process. For example, in meteorological research, weather stations distributed around the world collect meteorological data such as temperature, humidity, wind speed, and direction in real time through equipment such as temperature sensors, humidity sensors, and wind speed sensors. In biological experiments, microscopes, centrifuges, gene sequencers, and other instruments are used to observe cell structure, isolate biological samples, determine gene sequences, etc., and produce a large number of images, numerals, and text data. These data are typically structured numerical data with a clear physical significance and unit of measurement, but the accuracy, range of measurement, and frequency of sampling may vary from instrument to instrument, resulting in different data quality and reliability (Johnson & Roussou, 2022).

Analog data are generated by computer simulation and modeling. When certain scientific phenomena cannot be experimented with because of the high cost, difficulty in controlling experimental conditions, or long experimental cycles, scientists use computer simulations to reproduce these phenomena. For example, in astronomy, computer simulations of the formation and evolution of galaxies are used to study the birth, evolution, and properties of black holes; in materials science, molecular dynamics simulation is used to study the motion and interaction of atoms in a material and to predict the properties of a material. Analog data are usually generated based on physical equations and mathematical models and stored in the form of multi-dimensional arrays or grids, which reflect the state and change of simulated objects under different conditions. However, the accuracy of simulation results depends on the models and assumptions used. Different models may produce different simulation results, so it is necessary to verify and evaluate the analog data.

Text and literature data are also important parts of scientific data. Text materials such as scientific papers, experimental reports, and patent documents contain a large amount of scientific knowledge and research results. These data exist in unstructured text form but also contain some semi-structured data, such as tables and graphs. Text data contain information such as scientists' research ideas, experimental methods, observations, and conclusions and are of great value for understanding the progress and trends of scientific research. However, the processing of text data is relatively complex and requires the use of natural language processing techniques to extract critical information and knowledge.

In addition, scientific research often requires the integration of multiple modalities of data to provide a comprehensive and in-depth understanding of scientific phenomena. For example, functional magnetic resonance imaging (fMRI) provides structural and functional information of the brain, electroencephalography (EEG) records of electrical activity in the brain, and gene expression data to study brain function and disease mechanisms. In environmental science, it is necessary to consider the macro-environmental information provided by satellite remote sensing images, the micro-environmental data monitored by ground sensor networks, and the public's perception and feedback of environmental issues through social media. Integration of multi-source heterogeneous data is a challenging task because data from different sources may have different formats, scales, and semantics, requiring data preprocessing and transformation to achieve data compatibility and uniformity. Data visualization can help scientists better understand the relationships and differences between different sources of data by presenting them in the form of intuitive graphs, images, and interactive interfaces, so they can more effectively integrate and utilize these data.

1.3.2. *High-Dimensional Dynamics: Depth and Fluidity of Scientific Data*

In addition to multi-source heterogeneity, scientific data are characterized by high-dimensional dynamics, i.e., the data contain many variables, and these variables change with time and space (Liu & Chen, 2021).

High-dimensional data refer to a collection of data with many variables. In modern scientific research, with the continuous development of measurement technology and the increasing complexity of research issues, scientists need to consider multiple factors and variables at the same time to gain a comprehensive and in-depth understanding of the subject. For example, in the study of genomics, single-cell RNA sequencing technology can simultaneously detect the expression levels of tens of thousands of genes in a cell, forming a matrix containing a large number of genomics expression data; in materials science, the study of the properties of materials needs to consider multiple variables such as the chemical composition of the material, crystal structure, and process parameters. The number of these variables may reach dozens or even hundreds. In climate science, global climate models need to take into account dozens of variables such as atmospheric temperature, humidity, air pressure, wind speed, ocean temperature, and salinity, and these variables have corresponding values at different geographic locations and points of time. The complexity of high-dimensional data is that there can be complex nonlinear relationships between variables that can be difficult to observe and understand in an intuitive way. Human cognitive abilities are limited when dealing with high-dimensional data, and it is difficult to directly discover potential associations and patterns between variables in high-dimensional information. Data visualization techniques can project high-dimensional data into low-dimensional space by means of dimensionality reduction, such as principal component analysis (PCA) and t-SNE. It is also possible to demonstrate the values of multiple variables at the same time through visual methods such as parallel coordinates and radar graphs, helping scientists discover correlations and patterns between variables (Sacha *et al.*, 2020).

Dynamic data refer to data that change over time or space. Many scientific phenomena are dynamic, such as the motion of celestial bodies, the growth and development of living things, the evolution of the climate, and the process of chemical reactions. These dynamic processes need to be recorded through continuous or intermittent observations and measurements, resulting in a large amount of dynamic data. For example, in astronomy, the evolution of stars can be studied through long-term observation of parameters such as the luminosity and spectra of stars; in biology, it is possible to understand the developmental mechanism of a living organism by monitoring the proliferation, differentiation, and migration of cells in the body; in environmental science, through real-time monitoring of atmospheric pollution, water quality indicators, and other parameters, we can grasp the dynamic changes of environmental quality. The processing of dynamic data requires consideration of time series analysis and spatial analysis methods to reveal the patterns of data change over time and space. Data visualization can show dynamic data in an intuitive way through timeline animation, dynamic charts, etc., helping scientists observe the changing trends, periodicity, mutation points, etc. of the data, so as to better understand the inner mechanism of dynamic processes.

The combination of high-dimensional dynamic data adds further complexity to scientific data. Scientists not only deal with many variables but also consider how these variables change over time and space. For example, in weather forecasting, changes in multiple weather variables (such as temperature, humidity, air pressure, and wind speed) at different geographic locations and time points need to be considered at the same time to predict future weather changes; in biomedical research, it is necessary to analyze the dynamic changes of high-dimensional data such as gene expression data and proteomics data in the development of diseases to find diagnostic markers and therapeutic targets of diseases. Data visualization can show high-dimensional dynamic data in intuitive graphics and animation form through multi-dimensional dynamic visualization technology, helping scientists to observe and analyze data at the same time on multiple dimensions, thereby gaining a deeper understanding of the nature and laws of scientific phenomena.

1.3.3. *Uncertainty: The Blurred Boundaries of Scientific Data*

Scientific data are not absolutely accurate and reliable, but a certain amount of uncertainty exists. This uncertainty is mainly due to measurement errors, data missing and noise, and model uncertainty.

Measurement error is one of the common sources of uncertainty in scientific data. In practice, the result of a measurement is always different from the real one because of the limitations of the accuracy of the measuring apparatus, the influence of the measuring circumstance, and the error of the measuring personnel. For example, when using ordinary thermometers, the minimum scale may be 1°C, which means that the result of a measurement can only be accurate to 1°C, and there is ±0.5°C error. The result of a measurement may be less accurate because of the sensitivity and stability of the balance. The presence of measurement errors affects the accuracy and reliability of data, which in turn affects scientists' understanding of scientific phenomena and conclusions (Kitchin, 2022).

Data loss and noise are also common problems in scientific data. In the actual data collection process, some data may be missing due to sensor failure, experimental disruption, data transmission errors, etc. For example, in a long period of ecological environment monitoring, the sensor may be out of order because of battery exhaustion, bad weather, and other reasons, resulting in some periods of data acquisition that cannot be normal; in medical data records, some medical record information may be missing as a result of patient neglect or recording errors. At the same time, the data may also contain noise, i.e., random interference signals, which mask the true characteristics and patterns of the data. For example, in the measurement of electronic signals, due to electromagnetic interference, circuit noise, etc., the measured signal may contain a large amount of random noise, which affects the analysis and interpretation of the signal (Tversky & Morrison, 2021).

Model uncertainty refers to the uncertainty of data based on model simulations and predictions due to the limitations of scientific models and unreasonable assumptions. In scientific research, scientists often need to build mathematical models to describe and predict scientific phenomena. However, these models tend to simplify and approximate complex realities, ignoring certain secondary and complex factors. For example, in climate models, complex processes such as cloud formation and evolution and the interaction between the oceans and the atmosphere may be simplified because these factors are not sufficiently understood, resulting in some bias in model predictions. In addition, there is uncertainty in the parameter estimation of the model, and different parameter values may cause differences in the output of the model.

Uncertainty poses great challenges for the analysis and interpretation of scientific data. Scientists need to consider the impact of uncertainty and evaluate the reliability and credibility of data and conclusions. However, traditional data presentation methods such as tables, formulas, and static charts often struggle to visually express the uncertainty of the data, making it difficult for scientists to fully consider the factors of uncertainty when analyzing and making decisions. It can help scientists to understand the characteristics and rules of data and make scientific decision-making more reasonable.

The complexity of scientific data, including multi-source heterogeneity, high-dimensional dynamics, and uncertainty, poses many challenges to scientific research. As a powerful tool, it can present the complex data in an intuitive and understandable way and help scientists to understand the data, discover the rules, and solve the problem in science. In the following sections, we delve into how data visualization can provide effective solutions to these complexities.

1.4. Human Cognitive Limitations and The Need for Visualization

Although scientific data contain a wealth of information that can provide a key basis for revealing the laws of nature and advancing scientific and technological progress, humans face many constraints in understanding these complex data. The data generated in scientific research are often large, complex, and abstract, which makes them difficult for humans to extract valuable information directly and efficiently based on their own cognitive abilities (Ware, 2020). As a powerful tool, data visualization can transform abstract data into intuitive visual form, effectively compensating for the lack of human cognition and helping us better understand and explore the scientific world. This chapter discusses the limitations of human cognition and the necessity of visualization from three aspects: the advantage of visual cortex, the instinct of pattern recognition, and the management of cognitive load.

1.4.1. *Advantages of the Visual Cortex*

The human brain is a very complex and sophisticated organ, in which the visual cortex plays a crucial role in information processing. The visual cortex is the area of the brain devoted to processing visual information, which occupies a considerable proportion of the cerebral cortex, with an extremely large number of neurons and a complex network of neural connections. This makes human beings have strong visual information processing capabilities, and they can quickly and efficiently receive and interpret a lot of visual information.

Compared to other senses, vision has unique advantages. As a visual organ, the eye can receive a large amount of light information per second and convert it into neural signals to be transmitted to the brain. The visual cortex can rapidly process and analyze these signals, allowing us to perceive characteristics such as shape, color, size, position, and movement of objects in an instant. For example, when we walk into a room, our vision system can instantly capture the position and status of the various furnishings, decorations, and people in the room without complex thinking and reasoning.

Visual cortex also excels when it comes to processing complex information. It can process and integrate the received visual information in a layered manner, quickly identifying patterns, structures, and relationships in the information. For example, when we look at a map, the visual cortex can quickly identify the location and distribution of geographic elements, such as mountains, rivers, and cities, and understand their relative relationships. This powerful visual information processing capacity enables human beings to quickly obtain a lot of information through vision, which provides important support for our cognitive and decision-making.

In the process of cognition of scientific data, the advantages of the visual cortex are particularly evident. Scientific data often exists in various forms, such as numbers, tables, and graphs, but these forms are often difficult to understand and process directly by the human cognitive system. Through data visualization technology, scientific data can be transformed into visual forms such as graphs, images, and animations, which can fully utilize the advantages of the visual cortex, enabling scientists and researchers to obtain information from data more quickly and intuitively. For example, by drawing a visual graph of weather data, we can visually see the distribution and trends of weather factors such as temperature, air pressure, and wind speed over time and space without spending a lot of time analyzing and interpreting complex numerical tables.

1.4.2. *Instinct for Pattern Recognition*

Pattern recognition is one of the basic abilities in the course of human cognition, and it is also a kind of instinct formed during human evolution. Humans are born with the ability to recognize patterns in our surroundings. This ability allows us to quickly understand and adapt to the world around us. From simple geometric patterns to complex natural phenomena, from daily life to scientific data, human beings can quickly find patterns and rules by means of pattern recognition.

In our daily life, examples of pattern recognition are everywhere. When we see thick clouds in the sky, we instinctively realize that it may be about to rain; when we hear the call of birds, we can according to the pattern recognition of different birds; when we observe a group of people walking posture, we can based on their steps and movement pattern recognition of their mood and intention. These examples fully demonstrate that human beings have a powerful capacity for pattern recognition and can quickly extract significant patterns from complex information.

Pattern recognition is also of great significance in scientific study. An important goal of scientific research is to discover hidden patterns and laws from large amounts of data, thus revealing the nature and inner mechanisms of natural phenomena. However, scientific data are often complex, diverse, and massive, which contains a lot of noise and noise, making pattern recognition very difficult. Data visualization technology can present scientific data in an intuitive visual form, highlighting patterns in the data through graphics, images, and animations and helping scientists and researchers more easily identify and discover these patterns.

In astronomy, e.g., by visualizing large amounts of galaxy observation data, scientists have discovered that galaxies are distributed in certain patterns, such as the presence of clusters and filament structures. The discovery of these patterns has important implications for our understanding of the structure and evolution of the universe. In astrophysics research, through the visualization of stellar evolution data, scientists can more intuitively observe the patterns and characteristics of the evolution of stars at different stages, thereby deepening the study of the processes of star formation, evolution, and death. Data visualization stimulates humans' instinct of pattern recognition by transforming abstract data into intuitive visual form, which enables us to discover patterns and rules in scientific data more quickly and accurately, and promotes the progress of scientific research.

1.4.3. *Management of Cognitive Load*

Human cognitive resources are finite, and our brains have a certain limitation when it comes to processing information. When faced with a lot of complex information, our cognitive system will feel stress, resulting in information processing efficiency, and even cognitive fatigue and errors. Cognitive load management (CLM) is the process of organizing and presenting information in a reasonable way to reduce human cognitive burden and improve the efficiency and accuracy of information processing (Ware, 2020).

In scientific research, scientists and researchers often need to deal with huge amounts of complex data, which may come from multiple different data sources, in different formats and structures. For example, in genome research, a genome dataset may contain 1 billion base pairs of information but also may contain many gene expression data, protein structure data, and so on. When processing these data, scientists need to consider multiple variables and factors at the same time, analyzing their interrelationships and effects. If the data are presented to scientists in raw, unprocessed form, e.g., many digital tables and text files require scientists to spend a lot of time and effort sifting through, organizing, and analyzing the data, which not only increases the cognitive load but also tends to lead to information omissions and errors (Sacha *et al.*, 2020).

Data visualizations can effectively manage cognitive load by presenting complex data in a concise and intuitive visual form, reducing the amount of information scientists have to process and highlighting the key information and important patterns in the data. For example, by drawing a visual heat map of gene expression data, scientists can visually see differences in the expression levels of different genes in different versions, quickly identify high- and low-expression genes, as well as associations between gene expression patterns and disease or other biological phenomena. This intuitive visual presentation can help scientists understand data more quickly, reduce cognitive load, and improve research efficiency.

In addition, data visualization can further reduce the cognitive load by being interactively designed to allow scientists to explore and sift through data according to their own needs and interests. For example, in a data visualization interface, scientists can select areas of data they are interested in, view detailed data information, and adjust the presentation and parameters of the data. This interactive and visual approach allows scientists to actively participate in the process of data exploration, according to their cognitive abilities and needs, and flexible management of cognitive load, and improve the accuracy and efficiency of information processing.

Human beings face many limitations in recognizing scientific data, such as pattern recognition difficulty, cognitive overload, and so on. Data visualization takes full advantage of the human visual cortex. Stimulating the instinct of pattern recognition and effectively managing cognitive load can help us better understand and process complex scientific data, reveal the laws and information behind the data, and provide strong support for scientific research and decision-making. In today's data-driven science era, data visualization has become an indispensable tool in scientific research, and it will continue to play an important role in driving the continuous progress and development of science.

1.5. The Intervention Value of Digital Media: Making Up for the One-Way Transmission of Traditional Scientific Charts with Dynamic Graphics, Interactive Experiences, and Immersive Narratives

In the long process of scientific discovery, data visualization as a bridge between data and cognition has become more important as scientific paradigms have evolved. However, in the face of the complexity of contemporary scientific data, traditional science charts have gradually revealed limitations such as one-way transmission, lack of interactivity, and immersion. The intervention of digital media has brought revolutionary changes to scientific visualization. Through dynamic graphics, interactive experience, and immersive narrative, it has greatly compensated for the shortcomings of traditional scientific graphics and become the key force to promote scientific discovery and cognitive expansion.

1.5.1. *Limitations of the One-Way Transmission of Traditional Scientific Charts*

Traditional scientific charts, such as column charts, line charts, and pie charts, are classic forms of scientific data presentation. They have played an important role in scientific research and information exchange by presenting the basic features and trends of data in concise visual language. However, these charts are primarily a one-way messaging tool designed to present processed and simplified scientific data to an audience in an intuitive way, allowing audiences to quickly access the critical information in the data (Johnson & Roussou, 2022).

In the traditional model of scientific research, scientists obtain data through experimentation or theoretical derivation and then organize the data into a chart to present the results to their peers or the public. This delivery is often one-way, from the scientist to the audience, who receive information more passively and lack the opportunity for active exploration and in-depth understanding of the data. In academic papers, e.g., graphs are often used as supplementary notes to the findings of the study, and readers can only understand the data based on the information and annotations already in the graphs, making it difficult to further mine and analyze the data according to their own needs and interests.

In addition, traditional scientific charts are relatively homogeneous in their presentation form, mainly consisting of static two-dimensional graphs, and have limited ability to display complex multi-dimensional data, dynamic processes, and space-time changes. In the face of massive, multi-source, heterogeneous modern scientific data, traditional scientific charts often fail to fully and accurately present the rich contents and interrelationships of the data and are difficult to meet the needs of scientists and the public for in-depth exploration and understanding of scientific data.

1.5.2. *The Advantages of Dynamic Graphics Brought About by Digital Media Intervention*

The development of digital media brings the powerful function of dynamic graphics to scientific visualization. Dynamic graphics can present scientific data in a dynamic way, revealing the trends and dynamic processes of data over time through the presentation of time series. This dynamic presentation makes scientific data more vivid and intuitive and can help scientists and the public better understand the nature and patterns of scientific phenomena (Skarbez *et al.*, 2020).

For example, in astronomical research, dynamic graphs can show the evolution of galaxies, the formation and evolution of stars, and the motion trajectory of planets. Scientists can turn long-scale astronomical observations into dynamic videos or animations that allow viewers to visualize the formation and evolution of galaxies and the birth, growth, and death of stars. This dynamic display can not only help scientists better understand the evolutionary mechanism of astronomical phenomena but also popularize astronomical knowledge to the public and stimulate the public's interest and curiosity in astronomy.

In the field of biology, dynamic graphics can be used to show the movement of biological molecules, cell division and differentiation, signal transmission of biological neural networks, etc. Through dynamic simulation and visualization, scientists can observe the dynamic changes of biomolecules within cells and understand the mechanisms of various biological processes within a cell. For example, by showing protein folding through dynamic graphics, scientists can visually see the process of proteins gradually folding from disordered chains of amino acids into three-dimensional structures with specific functions, which have important implications for understanding protein functions and the mechanisms underlying disease.

1.5.3. *Interactive Experiences: Making the Audience a Participant in Scientific Exploration*

Digital media technology introduces rich interactive experience to scientific visualization so that the audience is no longer a passive recipient but can actively participate in the exploration and analysis of scientific data. Through interactive design, such as mouse clicking, dragging, zooming, and rotating, audiences can observe and analyze scientific data from multiple perspectives and levels according to their needs and interests.

Interactive scientific visualization devices have become a common form of display in science museums and science exhibitions. For example, through the touch screen, gesture recognition, and other views of technology, viewers can interact with scientific visualization content and explore the mysteries of scientific phenomena. In an interactive exhibition about the Earth's climate change, the audience can touch screen to select different time scales and geographical areas, observing the global temperature, sea level rise, melting glaciers, and other climate change indicators. Audiences can also simulate different climate change scenarios and learn about the impact of human activities on climate change by using actions such as dragging a slide (Johnson & Roussou, 2022).

In the field of scientific research, interactive scientific visualization provides scientists with more convenient and efficient means. Scientists can carry out real-time processing and analysis of scientific data and explore the potential patterns and patterns in the data. For example, in genomics research, scientists can use interactive visualization tools to interactively analyze gene expression data, protein structure data, etc., and discover interaction relationships and regulatory mechanisms between genes.

1.5.4. *Immersive Narratives: Creating Immersive Science Experiences*

Immersive narrative is another important embodiment of digital media in scientific visualization. Through virtual reality (VR), augmented reality (AR), mixed reality (MR), and other sciences of technology, visual science can create an immersive experience (Roussou, 2022).

In scientific research, immersive visualization can help scientists gain a deeper understanding of scientific phenomena. For example, in the study of physical science, scientists can use VR technology to observe the formation and evolution of galaxies and feel the vastness and mystery of the universe. In geological research, scientists can use AR technology to superimpose information such as geological formations and rock features in real scenarios during field trips, to understand the geological evolution process more intuitively.

In the field of science education and science popularization, immersive narratives can allow the public to experience scientific phenomena more immersive and stimulate the public's interest and love of science. For example, through VR technology, the public can immerse themselves in places such as the space station, the deep-sea floor world, and the interior of volcanoes, to learn about the latest scientific research and the mysteries of scientific phenomena. In the popular science exhibition, through AR technology, the public can scan the exhibits with a mobile phone or tablet to obtain more abundant scientific information and interactive experience, enhancing the fun and effectiveness of popular science education.

1.5.5. *The Role of Digital Media Interventions in Advancing Scientific Discovery*

The intervention of digital media, through dynamic graphics, interactive experiences, and immersive narratives not only compensates for the one-way transmission of traditional scientific charts but also brings new opportunities and impetus to scientific discovery. It makes scientific data more vivid, intuitive, and interactive and can help scientists understand scientific phenomena more deeply and discover potential patterns and patterns in the data. At the same time, it can more effectively disseminate scientific knowledge to the public, stimulate public interest and curiosity in science, and promote the popularization and dissemination of science culture.

Under the professional background of the School of Digital Media and Creative Technology, cultivating talents with the abilities of digital media technology and scientific visualization will help promote the innovation and development of scientific visualization technology and provide more powerful support for scientific research and scientific communication. By combining the creative expression of digital media with the rigorous logic of scientific discovery, we can open a new chapter in scientific understanding and allow scientific data to shine brighter with the help of digital media (Tversky & Morrison, 2021).

Chapter 2

Foundational Theories of Data Visualization and Digital Media Language

2.1. Introduction: When Scientific Data Meets Digital Media Languages — The "Translator" and "Magnifier" of Visual Theory

In scientific exploration in the digital age, data have become the core carrier of the "fourth paradigm" after experiments, theory, and calculations — from the recording of trillions of collision events per second by particle colliders to satellite remote sensing of surface observations at the pixel level per square kilometer. From the sequence of 1 billion base pairs from the gene sequencer to the tens of millions of environmental parameters collected in real time by smart sensor networks. These data are like "stars" scattered in the digital universe, containing natural laws and scientific mysteries, but due to their high dimensions, complexity, and abstraction, they are difficult to be directly perceived by human intuition (Kitchin, 2022).

At this point, the digital media language became a key bridge between data and cognition: It was a set of visual symbols (such as color). A "translation system" of shapes, lines, interaction logic (e.g., click, zoom, and filter), and narrative structure (e.g., contrast, association, and chronology) can turn cold numbers and symbols into intuitive visual images (e.g., a thermographic showing temperature distribution). Furthermore, through dynamic interactions (such as parameter adjustment, real-time observation, and multi-view linkage exploration), humans' instinct of pattern recognition is activated, and finally, the leap from "data fragmentation" to "scientific insight" is realized (Ware, 2020).

The foundational theory of data visualization is the digital media language's "syntax manual" and "design guide" — it not only answers technical questions about how to transform data into visual expression (such as which color to map numerical sizes and what kind of graphics are used to encode and classify variables) but also reveals the scientific principles of "why this transformation can effectively promote cognition" (such as the sensitivity of the human vision system to color contrast and the laws of gestalt psychology to visual organization). It can be said that without solid foundational theoretical support, digital media language will become disorderly "visual noise," and without the flexible application of digital media languages, basic theory will be stuck in abstract "paper talk."

2.2. Classical Visualization Theory: The Evolution from Cognitive Framework to Methodological System

Data visualization is not a simple "data mapping" but a scientific practice based on a solid theoretical foundation. Classical visualization theory provides a core framework for understanding "how to transform data into effective visual expression," guiding the goal setting, method selection, and effect evaluation of visual design. These theories start from the cognitive mechanism of human beings, combined with the needs of information science, graphics, and domain science, and gradually form a complete knowledge framework covering cognitive principles, design principles, interaction logic, and methodological systems. This section systematically sorts out the classical theories in the field of data visualization, deeply analyzes their core views, development context, and practical value, and lays the theoretical foundation for the application and innovation of visualization technology in the following chapters.

2.2.1. *Cognitive-Driven Theory: The Underlying Logic of Why Visualization Works*

The reason why visualization can improve the efficiency of human understanding of data is essentially because it fits the cognitive characteristics of the human visual system — the human brain processes visual information much faster than texts or numbers (studies show that about 80% of human external information is obtained through vision and can recognize simple visual patterns within a few hundred milliseconds). The core of cognitive-driven theory is to map abstract data into visual features (such as location, color, and size) that can be intuitively perceived by designing graphic coding that conforms to the laws of visual perception, thereby activating the brain's pattern recognition ability and accelerating information extraction and reasoning (Ware, 2020).

1. The Physiological and Psychological Basis of Visual Perception
There are significant differences in the sensitivity of human visual systems to different visual channels, which is the biological basis for visual coding design. Psychophysical scaling proposed by psychologist Stanley Smith Stevens and later scholars' classification studies of visual channels (such as Mackinlay's classical quantitative research) show the following:

- **High-precision channel:** Position (especially the one-dimensional position along the coordinate axis) and length (such as the length of a bar graph) have the highest resolution accuracy for numerical differences (error rate <5%), suitable for showing accurate comparison of quantitative data (e.g., "Region A's GDP is 20% higher than Region B").
- **Medium precision channel:** This includes medium resolution accuracy (error rate of about 10–20%) for angles (such as the fan angle of a pie chart) and area (such as the size of bubbles in a bubble chart), which can be used to show the relative ratio or distinction of secondary variables (such as "market share share of three types of products").
- **Low-precision channels:** Weak perception of subtle differences such as color saturation, texture density, and curve curvature (error rate >30%) is more suitable for encoding categorical variables (e.g., "different countries are differentiated by different colors") or assisting in highlighting key points (e.g., highlighting anomalous data in red).

In addition, the principles of visual organization proposed by gestalt psychology (e.g., proximity, similarity, closure, and continuity) further reveal how humans spontaneously combine discrete visual elements into meaningful wholes — e.g., similar points are considered the same category, areas of the same color are associated with related data, and continuous lines are understood as trend changes. These principles provide direct guidance for "visual metaphorical design" in visualizations (Sacha *et al.*, 2020) (e.g., clustered clusters of points for clustering of data and gradient color bars for the distribution of continuous variables).

2.2.2. *Cognitive Load Theory and Visualization Efficiency*

The cognitive load theory proposed by Australian educational psychologist John Sweller points out that human working memory capacity is limited (usually only 4±1 information blocks can be processed at the same time), and when visual design introduces too much redundant information (such as complex decorative elements and too many color classifications) or requires users to perform difficult cognitive transformations (such as reverse reasoning of unintuitive color mappings into values), it will add an additional "extraneous" (cognitive load), which reduces the efficiency of information acquisition (Sweller, 2020).

Visual design optimizes cognitive workload by the following:

- **Reduce redundancy:** Retain only visual elements relevant to the core scientific question (e.g., avoid adding irrelevant border decorations to heat maps showing temperature distributions).
- **Match cognitive habits:** Use familiar visual coding (e.g., "rising arrow" for growth and "red" for danger) or high temperature, reducing the cost of semantic conversion.
- **Hierarchical presentation:** Dynamically adjust the information density through interactive technologies (such as zooming and filtering) (such as the initial view shows the global trend and the user clicks to expand the local details) to avoid presenting excessive information at once.

2.2.3. *Task-Data-Coding Theory: A Methodological Framework for Visual Design*

Cognitive-driven theory explains "why visualization works," while Task-Data-Encoding Framework further answers "how to design effective visualization" — it regards visualization as a system engineering of "processing specific data for a specific task and achieving efficient cognition through optimal coding," clarifying the three core elements in the design process and their related logic.

1. The Three Core Elements of Theory

- **Data type:** The nature of scientific data determines the encoding methods available. Common data types include the following:
  - **Quantitative:** This is numerical and sortable (e.g., temperature, GDP, and time), usually encoded with high-precision channels such as position (coordinate axis), length (bar graph), and area (bubble graph).
  - **Categorical data:** This represents discrete categories or groupings (such as countries, experimental conditions, and species types), commonly used colors (different hues), shapes (different icons), textures (different patterns), and other low-precision channels.
  - **Temporal data:** Data with time dimension (such as stock price trends, meteorological observation records) should reflect the time logic through the timeline (x-axis), animation (dynamic change), or sequence arrangement (such as the chronological order of line charts).
  - **Spatial:** Data related to geographical or geometric locations (such as population distribution maps and earthquake epicenter locations) should be displayed through spatial reference systems such as map projections (such as Mercator projections) and three-dimensional terrain.
- **Visualization task:** The specific goal of the user using visualization can be divided into basic tasks and advanced tasks:
  - **Basic tasks:** These include data retrieval (e.g., "Find the data point with the largest value"), comparison (e.g., "Which is higher, A or B?"), and distribution analysis (e.g., "What is the concentration trend and discreteness of the data?").
  - **Advanced tasks:** These include tasks related to pattern discovery (e.g., "Is there periodicity or clustering in the data?"), relational reasoning (e.g., "Is there a correlation between variables X and Y?"), and causal inference (e.g., "How do parameter adjustments affect the final result?").
- **Visual encoding:** Mapping data attributes to visual channels (e.g., "temperature values in y-axis coordinates" and "population density in color shades"). Effective coding needs to meet three principles:

accuracy (the visual characteristics after encoding correspond strictly to the original data), discernibility (users can easily distinguish different encoding values), and memorability (users can recall coding rules for a long time).

2. Application Process of Theory: Mapping from Requirements to Design

Visual design guided by task-data-coding theory is an iterative process of "requirements analysis → data abstraction → task definition → coding selection → evaluation and optimization":

- **Define user tasks:** Identify the core goals of visualization (e.g., "verify the accuracy of climate model predictions" or "show the public the global distribution of carbon emissions") by communicating with scientists, engineers, or public users.
- **Analyze data characteristics:** Identify the type (quantitative/categorized/temporal/spatial), dimension (univariate/multivariate), scale (small sample/massive data), and constraints (such as the accuracy of physical units and privacy protection requirements).
- **Decompose specific tasks:** Break down user goals into actionable subtasks (e.g., "compare temperature averages in different years" and "identify the correlation between temperature and precipitation").
- **Select the optimal encoding:** Match the most suitable visual channel according to the data type and task requirements (e.g., when displaying multivariate timing data, the x-axis can represent the time, the y-axis represents the temperature value, and the line color represents the precipitation level).
- **Evaluate and iterate:** Verify the validity of the coding through user testing (e.g., eye tracking, task completion time measurement) and adjust unreasonable mappings (e.g., avoiding distinguishing important categories with similar colors).

2.2.4. *Visual Methodology System: The Evolution from Static Charts to Interactive Exploration*

With the increase in data complexity (such as multi-dimensional, dynamic, and large-scale data), the classic visualization methodology has gradually developed from "static single-view display" to "dynamic multi-view interactive exploration," forming a method system covering univariate display, multivariate association, spatial–temporal analysis, high-dimensional dimensionality reduction, and interaction design.

1. Static Visualization Method: Cognitive Optimization of Basic Charts

Static charts are the earliest and most widely used form of visualization, and their classic types include the following:

- **Univariate display:** These are histograms (display numerical distribution), pie charts (display classification ratio), and line charts (display trend change).
- **Bivariate associations:** These are scatter plots (showing the correlation between two quantitative variables), bar charts (comparing quantitative values of different classifications), and matrix tables (showing frequencies or values between categorical variables).
- **Multivariate synthesis:** These are radar plots (showing the overall outline of multiple quantitative variables), stacked plots (breaking down the classification of totals), and small multiples plots (comparing data under different conditions through multiple similar subplots) (Tufte, 1983).

Modern static visualization theory emphasizes the balance between "visual simplicity" and "information density" — e.g., optimizing chart design through the "Data-Ink Ratio" (proposed by Edward Taft): removing unnecessary grid lines, borders, and decorative elements, and retaining only the "data ink" that carries the core data (such as lines and color blocks in the chart) to make key information more prominent.

2. Dynamic and Interactive Visualization Methods: From passive viewing to active exploration

In the face of large-scale or high-dimensional data (such as million-level sensor records and 10-dimensional scientific parameters), static charts are difficult to fully present all information, and dynamic and interactive technology has become the key. Its core approach includes the following:

- **Dynamic visualization:** This presents the dynamic process of data through timeline animations (e.g., showing interannual temperature changes) and data stream updates (e.g., real-time monitoring sensor data) to help users observe trends and event evolution.

- **Interactive exploration:** This provides interactive operations such as zoom, pan, filter, sort, and details-on-demand, allowing users to focus on specific data subsets according to their own needs (e.g., "only view epidemic data in a certain province" or "zoom in to observe an abnormal data point").
- **Linked views:** Multiple visual views (such as maps, line charts, and scatter charts) share the same data source, and when users select or manipulate in one view (such as clicking on an area on the map), other views synchronize highlighting or filtering related data (such as line charts showing the time series changes of that area) to assist users in establishing cross-dimensional data associations.

3. High-Dimensional Data Visualization: The Art of Dimensionality Reduction and Projection

When the data dimension exceeds 3D (such as gene expression data contain thousands of gene variables), humans cannot directly perceive the structure of high-dimensional space and need to map high-dimensional information to 2D or 3D space through dimensionality reduction technology (Maaten & Hinton, 2008; Inselberg, 2009). Classic methods include the following:

- **Linear dimensionality reduction:** Principal component analysis (PCA) projects high-dimensional data to the principal component directions with the largest variance through orthogonal transformation and retains the main variation information of the data.
- **Nonlinear dimensionality reduction:** *t*-SNE (t-distributed random neighborhood embedding) and UMAP (uniform manifold approximation and projection) are more suitable for capturing the local structure of data (such as the internal relationship of clusters) and are often used to visualize single-cell RNA sequencing data (showing the distribution patterns of different cell types).
- **Parallel coordinates:** Each dimension is represented as a parallel axis, and the data points are represented by a line connecting the corresponding values of each axis, and the association between high-dimensional variables is visually displayed through the intersection of the line and clustering (e.g., "gene combinations are strongly correlated with disease classification").

2.2.5. *Data-Ink Ratio Principle: Minimalism Focused on the Essence of Information*

Background and core ideas: This theory was first systematically proposed by American statistician Edward Tufte (1983) in his book "*The Visual Display of Quantitative Information*" and is known as the "golden rule of visual design." Taft observed that traditional charts are often filled with redundant elements (such as decorative borders, redundant ticks, and 3D effects), and these non-data-related "inks" not only distract users but can also obscure the true patterns of key data. Therefore, he proposed that the core goal of visual design is to maximize the "Data-Ink Ratio," i.e., to carry as much data information as possible with as little unnecessary ink (redundant visual elements) as possible.

Mathematical Expression and Practical Logic: Taft divides the total ink in the graph into two categories:

- **Data ink:** These are visual elements that are used directly to encode data values (such as lines in line charts, rectangular heights in bar charts, and point positions in scatter plots).
- **Non-data ink:** These are auxiliary or decorative elements (e.g., shadows outside the chart title, 3D stereoscopic effects on coordinate axes, and over-rendering of grid lines).

Data-ink ratio = data ink/(data ink + non-data ink)

Ideally, this ratio should approach 1 (i.e., almost all ink is used to pass the data itself). For example, compare two line charts: one with a gradient background, bold black border, and stereo axis, and the other with only the axis scales, clean gridlines, and clear polylines — the latter has a significantly higher data-to-ink ratio due to the exclusion of more than 90% of non-data ink, allowing users to identify trend changes faster.

Extended principles and typical cases: Taft further derives practical guidelines such as "minimizing non-data elements" (such as removing unnecessary grid lines and replacing dark scale lines with light gray) and "avoiding data redundant coding" (such as the same data are not repeatedly emphasized by color and shape). Typical applications include basic statistical charts in scientific papers (usually only coordinate axes, data lines, and key labels) and minimalist information dashboards (such as epidemic data daily reports only use bar charts + numbers to mark key indicators).

2.2.6. *Visual Coding Theory: Mapping Rules for Multi-dimensional Information*

Background and core framework: This theory was laid by French computer scientist Jacques Bertin (1967) in his classic book *Sémiologie Graphique* and was later developed and perfected by Stanford University professor Stephen Few and others. Starting from "how graphics convey information through visual variables," Bertin systematically defined the concept of visual encoding — i.e., mapping abstract data values into perceptible graphic elements through specific visual attributes (such as position, size, and color) to support rapid cognition of users.

Core visual variables and their semantic properties: Bertin divided visual variables into primary variables and secondary variables as follows:

Table 1. Visual primary variables and secondary variables by Bertin.

Variable Type	Specific Variable	Type of Data Encoded	Cognitive Properties
A type of variable	Position	Quantitative/ qualitative (such as coordinate axis values)	The most accurate encoding method. (Humans are most sensitive to spatial position, such as in scatter plots.) The position of the X/Y axis directly corresponds to the value.
	Size	Quantitative (e.g., numerical size)	It can distinguish relative differences (e.g., the radius of a circle in a bubble chart reflects the magnitude of data), but it is important to note the nonlinearity of the human eye's perception of area (usually judged by diameter rather than area).

Table 1. (*Continued*)

Variable Type	Specific Variable	Type of Data Encoded	Cognitive Properties
	Shape	Qualitative (e.g., category differentiation)	It is suitable for distinguishing scattered categories (e.g., different icons in different countries), but it is difficult to express numerical differences.
Class 2 variables	Color	Qualitative/ quantitative (e.g., classification/ gradient)	Hue (e.g., red/blue/green for different lines of business) and brightness/saturation for expressing numerical intensity (e.g., red from light to dark indicates increased risk level).
	Texture	Qualitative (e.g., category subdivision)	Distinguish similar categories by line density, dot matrix density, etc. (e.g., distinguishing rock strata types with different textures in geological maps).
	Orientation	Quantitative/ qualitative (such as angular changes)	It is rarely used alone and is often combined with position (such as the dimension corresponding to the sector direction in radar maps).
	Transparency	Quantitative (e.g., overlay hierarchy)	It reflects data weights through opacity (e.g., overlapping area transparency reduces display density).

Practical significance of the theory: The visual coding theory clarifies the matching rules of "data type-visual variable" — e.g., time series data are encoded with position (X-axis time + Y-axis value) to ensure accuracy; in multi-dimensional analysis, high-dimensional information compression can be achieved by combining variables (such as the position of the dots in the scatter plot = two quantitative variables, color = categorical variables, and size = third quantitative variable). Typical applications include weather data visualization (color gradients for temperature and contour positions for altitude) and business dashboards (bar chart height + color to differentiate sales areas from performance levels).

2.2.7. *Graphical Syntax Theory: Modular Diagram Construction Logic*

Background and core innovations: Traditional diagram design relies on designers' empirical intuition and lacks systematic generation rules. To solve this problem, statistician Leland Wilkinson (1999) proposed a formal graph generation framework. The Grammar of Graphics was later developed by Hadley Wickham and others into the popular R language visualization package ggplot2 ("gg" is the abbreviation for "Grammar of Graphics"). This theory regards graphs as a "language" composed of multiple semantic modules (such as data, mapping, coordinate systems, and geometric objects) according to rules, thus realizing a structured and reusable construction process from data to graphs.

Core modules and generation process: Graphical syntax breaks down the chart into the following key components (using the most basic scatter plot as an example):

1. **Data:** It is the original dataset to be visualized (such as a table containing three columns: "City," "Population," and "GDP").
2. **Aesthetic mapping:** It defines the correspondence of data variables to visual attributes (e.g., "City→ abscissa position, GDP" → ordinate location, "population" → the size or color of points).
3. **Geometric object:** It determines the basic shape of the chart (e.g., "points" form scatter charts, "lines" form line charts, and "columns" form bar charts).
4. **Scale:** It controls the value range and mapping rules of visual attributes (such as linear/logarithmic scale on the Y axis and color grading scheme).
5. **Coordinate system:** It defines the layout of the data space (such as Cartesian coordinate system, polar coordinate system, and map projection coordinate system).
6. **Statistical transformation:** It pre-processes raw data (e.g., calculated mean, binning density, and smooth curve).

7. **Faceting:** It groups data by dimension and generates sub-graphs (e.g., generates a multi-year comparison matrix by faceting by "year").

Revolutionary significance of the theory: Graphical syntax shifts chart design from "experience-driven" to "rule-driven" — users only need to clarify, "What data do I want to show?," "What visual variable is used to encode?," and "What chart types do I need?" to quickly generate visualizations that meet your needs through a combination of modules. For example, if you want to compare the price and sales of different brands of mobile phones, you just need to do the following:

- **Data:** mobile phone brand, price, sales table.
- **Mapping:** brand → color, price → X-axis, sales → Y-axis.
- **Geometry objects:** scatter.
- Other modules remain default (Cartesian coordinate system and linear scale).

The resulting scatter plot automatically presents the distribution patterns of each brand's data points, eliminating the need to manually adjust every detail. This theory has become the underlying design logic of modern visualization tools, such as Tableau, Power BI, and Python's Plotly.

2.2.8. *Practical Value of Theory and Contemporary Development*

Classical visualization theory is not only the cornerstone of academic research but also a key tool to guide the design of practical projects. For example, in climate science, scientists use task-data-code theory to break down the goal of "validating global warming" into sub-tasks such as "showing time-series changes in global average temperature over the past century" and "comparing differences in projections under different emission scenarios." In turn, a line chart (timeline + temperature values) and a multi-panel contrast chart (overlayed display of different scenarios) were chosen as the core visualization form; in medicine, cognitive load theory instructs physician-side visualization systems to avoid displaying too many physiological parameters at the same time (such as heart rate, blood pressure, and degree of saturation) but through the hierarchical interaction (showing the key indicators overview, click and then expand the detailed waveform) to reduce the cognitive burden of diagnostic.

Contemporary visualization theory is still evolving, and the integration of artificial intelligence, virtual reality, and other technologies has spawned new research directions. For example, "automated visualization generation" based on deep learning attempts to automatically identify data types and user tasks through algorithms and recommend optimal coding schemes; immersive visualization theory focuses on how to use three-dimensional space layout and natural interaction (such as gesture manipulation) in VR/AR environment to improve the efficiency of complex scientific data exploration. These developments have further expanded the boundaries of classical theory, keeping it consistently adapted to the frontiers of science and technology.

Classical visualization theory builds a "cognitive-design-method" framework for data visualization — it starts with human visual perception and cognitive laws. It clarifies the core relationship between data, tasks, and coding, provides a systematic methodology from static diagrams to interactive exploration, and iterates as technology advances. Understanding these theories is not only the first step in mastering visualization techniques but also a key prerequisite for designing an "effective, efficient, and easy-to-use" visualization solution. In subsequent chapters, we delve into the practical applications of digital media language, such as graphic symbols, color semantics, and interactive design, based on these theoretical foundations. Additionally, we explore how cutting-edge technologies such as AI and VR are driving further innovation in this field.

2.3. Creative Languages of Digital Media

Based on classical visualization theory providing methodological support for data presentation, the rapid development of digital media technology has given visualization richer expression dimensions — from static graphs to dynamic interactions. From two-dimensional planes to immersive spaces, digital media have not only changed the carrier form of information but also expanded the possibilities of data narratives through unique creative languages. This creative language is a fusion of technology and art, incorporating technical syntax (such as animation) specific to digital media. Interaction and multimodal convergence also inherit the aesthetic principles of traditional visual design (such as emotional empathy and metaphorical expression), with the goal of making data not only "information to be seen" but also "stories to be experienced." This section analyzes how digital media can creatively reconstruct the relationship between data and users in three core dimensions: dynamic visualization language, interactive perception language, and multimodal fusion language (Heer & Shneiderman, 2021; Tversky *et al.*, 2020).

2.3.1. *Dynamic Visual Language: Narrative Tension in Time Dimensions*

Core concepts and technical foundations: Animated visualization refers to the transformation of static data into a dynamic process display with time continuity by changing visual elements on a time series (such as displacement, shape change, and color gradient). Its technical foundation stems from the "frame sequence" nature of digital media — using the human visual temporal effect to form a smooth dynamic perception by continuously playing a series of small variations of images (24–60 frames per second). Compared with static graph, the key advantage of dynamic language is that it can express the changing rule and cause and effect relationship of data.

Key creative techniques and narrative logic: Creative expression of dynamic visualization is achieved mainly through the following methods:

- **The progressive presentation of time axis driven:** Combining data over a long period of time (e.g., ten-year GDP growth), one-hundred-year climate change is compressed into a few to tens of seconds of animation that controls the rhythm of key frames (e.g., accelerating the unimportant smooth period and slowing down the mutational nodes) and guides the user to perceive the cumulative effect of the trend. For example, NASA's Global Temperature Change Animation, which shows surface temperature data from 1880 to 2020 in a color-gradient animation, allows users to directly see the process of red (hot-temperature abnormalities) spreading from local to global, which is more impactful than a static line chart (NASA, 2023).

- **The causal metaphor of state transformation:** This implies the logical relationship behind the data through morphological changes in elements (such as scatter points aggregated into clusters and lines connected into networks). For example, in showing the evolution of social network user relationships, isolated nodes in the initial frame are gradually connected by lines into communities over time, a dynamic process itself that is a metaphor for the social phenomenon of "interpersonal relationships moving from isolation to aggregation."
- **Path visualization of motion trajectories:** This reveals hidden dynamic patterns with continuous motion trajectory (e.g., particle flow and path tracking). Typical examples include superimposed animation of vehicle movements in traffic flow data (showing traffic hotspots during rush hour) and particle simulation of ocean current data (showcasing heat transport paths through the direction and speed of color particles flowing).

Practical value and user perception impact: The dynamic language reduces the cognitive threshold of complex time series data by the dual mechanism of "time slice and visual continuity" — users do not need to actively compare multiple static graphs; rather, the laws are naturally deduced by observing dynamic processes (such as the correlation between economic growth and rising carbon emissions). However, it is important to note that overly complex dynamic effects (such as meaningless rotations and flickering) can be distracting, so dynamic design needs to follow the principle of "motion for data narrative" (e.g., when Taft emphasizes that "animation should be a carrier of information, not a decoration").

2.3.2. *Interaction-Perception Language: User-led Exploration Experience*

Core concepts and technical foundations: Interactive perception language means dynamically adjusting the presentation and detail levels of data through real-time interactions between the user and the visual interface (e.g., clicking, dragging, zooming, and filtering) to support user-centric exploration. Its technical basis relies on the digital media input and output closed-loop system — sensors (mouse, touch screen, and eye tracker) to capture the user's behavior, the calculation engine's real-time response, and update the graphics rendering, forming a "operation-feedback-operation" dynamic cycle. Unlike the one-way transmission of traditional static charts, the essence of interactive languages is to return data exploration to the user, stimulating his ability to proactively discover hidden patterns.

Key interaction dimensions and creative strategies: The creative expression of interactive perception language is reflected in multi-dimensional interaction design:

- **Drill-down:** This allows users to freely switch the abstraction level of data through interaction (e.g., from national sales data to province, city, and store). For example, in a business intelligence dashboard, users click on a provincial column chart to expand a detailed curve of the respective cities, and interactive actions become a bridge between macro trends and micro details.
- **Multivariate association exploration:** This helps users discover potential associations between variables by cascading filtering (e.g., after selecting a color category, other charts synchronize highly highlighted relevant data) or slider tuning (e.g., dragging the Age Range slider to filter population data in real time). Typical examples include public health visualizations in which users observe in real time the impact of parameters such as "vaccination rates" and "age distribution" on the epidemic spread curve.
- **Dynamic manipulation of space and time:** In geographic information visualization (GIS), users can rotate a map's viewing angle by mouse-tapping, scrolling a wheel to zoom in on a local area, or tracing historical geographic states (such as satellite image superposition of a city's expansion) through a timeline slider. This interactive manipulation of the two dimensions of space and time makes abstract data tightly bound to the physical space of the real world.

User cognitive gains and design principles: Interaction languages reinforce the user's sense of cognitive engagement through "action feedback" — when the user actively screens out a certain class of anomalous data and observes corresponding graphic changes. Their understanding of data patterns is more profound (the "generative effect" in psychology: the knowledge acquired by active action is stronger). But it is important to balance "degrees of freedom" with "guidance": avoiding the loss of too many options (e.g., providing default views and common filter combinations) and reducing costs through visual cues (e.g., hover previews and tooltips). As Ben Shneiderman, an expert in interactive design, puts forward the classic principle: "Overview first, zoom and filter, then details-on-demand."

2.3.3. *Multimodal Blended Language: Immersive Communication Beyond Vision*

Core concepts and technical foundations: Multimodal fusion language refers to the integration of sensory channels other than vision (such as hearing, touch, and smell) or media forms (such as text, voice, and 3D models), through the co-coding of multiple dimensions of information, breaking the information-carrying limits of a single visual channel, and building a more three-dimensional data narrative scene. Its technological foundations stem from digital media cross-modal transformation technologies (such as mapping numerical values to sound frequency and generating 3D models of spatial data) and immersive device support (such as VR headsets, touch screens, and force feedback devices). In complex data scenarios (such as high-dimensional scientific data and abstract social indicators), multimodal language can effectively compensate for the limitations of a single visual dimension and improve the efficiency of information transmission and emotional resonance.

Key modal types and fusion strategies:

- **Audio:** Mapping data attributes to sound features (such as frequency, volume, and rhythm) to form "data sonification." For example, in the seismic monitoring system, the magnitude of the quake is mapped as the pitch of the sound, the depth of the quake is mapped to the pitch of the sound, and the user can quickly sense the temporal and spatial distribution of seismic activity by listening to the sound wave sequence. In financial markets, the rise and fall of stock prices can be distinguished by a short, high note (up) from a low, long note (down), to help traders capture momentary fluctuations.
- **Haptic:** It is the physical properties of analog data through vibration intensity, temperature variation, or force feedback. For example,

in medical training visualizations, the strength of contact between surgical instruments and human tissue is feedback through the vibration intensity of the handle (simulating real touch), helping learners understand operational precision requirements. In industrial equipment monitoring, the sensor vibration data can be converted to a slight vibration of the control panel to alert the operator to abnormal conditions.

- **3D/AR/VR (spatial):** It is the two-dimensional plane data extended to three-dimensional space or virtual environment, the use of depth perception and immersion to enhance cognition. For example, substructure data can be used to demonstrate interatomic bonding relationships through a 3D rotation model. In the visualization of urban planning, users wear VR equipment to "walk into" the future construction group and intuitively feel the spatial layout and flow density. Augmented reality (AR) overlays the data on the real scene (such as the distribution of underground pipelines after scanning a map) to achieve the combination of virtual and real information interaction.

The cognitive advantages of multimodal synergy: Different sensory channels are complementary — vision is good at processing spatial and structural information, hearing is sensitive to time series, touch provides direct physical feedback, and 3D space reinforces depth perception. Multimodal fusion, which leverages information redundancy and synergy across different channels (such as displaying bar charts simultaneously, playing sound cues corresponding to numerical values, and providing touchable 3D models), not only enhances the accuracy of information transmission (allowing users to verify the same conclusion through various means), but also evokes emotional resonance (e.g., using soothing music to accompany positive trends in environmental data, to increase public awareness of sustainability). However, it is important to note that multimodal design needs to follow the principle of "intermodal consistency" (e.g., the direction of the rhythm change of sound needs to be logically consistent with the direction of color gradient of vision) to avoid cognitive confusion resulting from channel conflicts.

Core values of creative languages for digital media: The creative language of digital media (dynamic, interactive, and multimodal) is essentially a technical extension of the "bandwidth" and "dimensions" of human perceived data — dynamic language allows time to "flow."

Interaction language allows the user to be the subject of exploration, while multimodal language breaks the limitations of a single sense. These creative languages are not a subversion of classical visualization theories, but rather complement and sublimate them: They guarantee data accuracy. Through more vivid, immersive, and personalized expressions, data can be transformed from "cold numbers" to "stories with temperature," ultimately driving the leap from "messaging" to "cognitive empathy." In subsequent chapters, we further explore how these creative languages can be combined with specific visualization techniques (such as dynamic charting libraries, interactive design frameworks, and mixed reality platforms) to achieve a theoretical-to-practical transition.

2.4. The Logic of Scientific Visualization and Digital Media

Scientific visualization is an important branch of data visualization. Focuses on transforming abstract data from natural science and engineering fields (such as physical distribution, biological structure, and fluid dynamics simulation results) into intuitive graphic images to support researchers' observation, analysis, and reasoning about complex phenomena. Traditional scientific visualization relies on static two-dimensional graphs or basic three-dimensional models, which can present key data characteristics. However, in the face of complex scenarios such as high-dimensional dynamic data (such as transient changes in turbulent fields, microprocesses of molecular motion) and multimodal data fusion (such as association analysis of medical imaging and genomic data), the limitations of insufficient expression are gradually revealed. The rise of digital media technology (including dynamic rendering, interactive awareness, and multimodal fusion) has provided new technical tools and creative language for scientific visualization. Through the deep integration of the two, a new paradigm of "data accuracy as the foundation, digital media creativity as the enabler" has been formed. This section analyzes the logic of scientific visualization and digital media from three aspects: the underlying motivation, the core mechanism, and the typical practice.

2.4.1. *Underlying Drivers of Convergence: The Two-Way Drive of Scientific Exploration Needs and Technological Capabilities*

The integration of scientific visualization and digital media is not accidental, but the inevitable result of the evolution of the scientific research paradigm and digital media technology.

From the scientific demand side, modern scientific research is shifting from "experience-driven" to "data-driven" — astronomical observations produce petabytes of galaxy spectral data, climate models output meteorological parameters per kilometer of global grid, and brain science studies obtain dynamic sequences of neural activity at millimeter-resolution through fMRI scans. These data are generally characterized by high dimensionality (multivariable coupling), high dynamics (continuous change in time/space), and high complexity (multi-physical field interaction). Traditional static charts (such as line charts and column charts) can only show sliced information in a single moment or dimension, making it difficult to show the correlation of global patterns of data with microscopic details. As scientists' cognitive needs have extended from "providing known hypotheses" to "discovering unknown patterns," more powerful visualization tools are urgently needed to support exploratory analysis of complex data.

From the technical supply side, breakthroughs in digital media technologies offer new possibilities for scientific visualization:

- **Computing performance improvements:** These include GPU parallel computing and cloud computing support real-time rendering of large-scale datasets (e.g., fluid simulations of hundreds of millions of particles).
- **Dynamic interaction technologies mature:** High refresh rate display screens and touch/gesture/eye movement interaction devices allow users to dynamically adjust data presentation through natural behaviors (e.g., zooming, rotating, and filtering).
- **Enhanced multimodal fusion:** This includes 3D modeling, virtual reality (VR), augmented reality (AR), and spatial audio technology, which can map abstract data into three-dimensional space or combine auditory/tactile feedback to break the limitation of two-dimensional expression.

- **Cross-platform collaboration:** Cloud-based visualization platforms allow global research teams to share dynamic analysis results in real time (such as co-evolution of disease transmission models).
- **The nature of two-way driving:** The need for deep interpretation of data in scientific research forces visualization technologies to break through the limitations of traditional static models. The toolset capabilities of digital media technologies provide an achievable solution to meet this need — a typical example of "demand-driven technology, technology enabling science."

2.4.2. *Core Mechanisms of Convergence: Multilevel Mapping from Data to Cognition*

The integration of scientific visualization and digital media is not a simple technical superposition but a hierarchical system of "data–code–presentation–interaction–cognition" to achieve efficient transformation from raw data to scientific insights. At the core of this mechanism is the use of the creative language of digital media (dynamic, interactive, and multimodal) to enhance the expressive accuracy and cognitive friendliness of scientific data, which can be divided into the following five key levels:

1. Data layer: Standardized access to multi-source heterogeneous data

Scientific data come from a wide variety of sources (e.g., sensor networks, simulation software, and experimental instruments), in a variety of formats (e.g., structured tables, unstructured point clouds, and multi-dimensional arrays), and are often accompanied by noise or missing values. Digital media technology first completes cleansing, alignment, and standardization through data preprocessing modules (such as unifying the physical quantities of different units into international standards and inserting discrete sampling data into continuous fields) and provides a unified "data interface" for subsequent visualization. For example, in climate science, satellite remote sensing data (infrared radiation values), ground-based weather station observations (temperature/humidity), and ocean buoyancy monitoring (current velocity) need to be integrated into consistent global grid data through space-time alignment technology to be used as visualization inputs.

2. Coding layer: Scientific semantics and accurate mapping of visual/ multimodal variables

Scientific data often have clear physical significance (such as the direction and size of the speed field, and bond length and bond angle of molecular structures), so visual coding needs to be strictly guided by the principles of scientific semantic accuracy — visual variables in digital media (such as color gradients, line thickness, and 3D model size) or non-visual variables (such as sound frequency, vibration intensity) must establish interpretable mapping relationships with key parameters of the data (such as numerical range, unit, and physical meaning). For example, in fluid mechanics visualization, the direction of the velocity vector is encoded by line orientation, and the size is encoded through line length or particle flow rate. In medical MRI data, tissue density differences are mapped by grayscale (or pseudo-color) to avoid color bias, leading to doctors misjudging the location of lesions.

3. The Presentation layer: Dynamic and multimodal enhancement of expression

In response to the high dynamic nature and complex structure of scientific data, digital media significantly improves the presentation effect through dynamic rendering and multimodal fusion:

- **Dynamic presentation:** This includes showing data evolution processes (such as the formation and dispersion of typhoons and molecular collision trajectories in chemical reactions) through time series animation, or revealing hidden laws through parameter tuning animation (such as changing the viscosity coefficient of a fluid and observing flow field morphology changes in real time).
- **Multimodal fusion:** Combine 3D models (such as a three-dimensional display of protein folding structures), spatial audio (such as the frequency and amplitude of seismic waves converted into low-frequency rumbles), and tactile feedback (such as simulating contact resistance between surgical instruments and tissues) to synchronize information from multiple sensory channels. For example, a geologist "enters" a fault model of the Earth's crust with a VR device, hears the sound effects of rock friction, and feels the vibration of the handle, perceiving fault activity far beyond a two-dimensional cross-section diagram.

4. Interaction layer: User-led exploratory analysis

The interactive capabilities of digital media give scientists the right to "proactively ask questions" — through interactive controls such as sliders to adjust the length of time steps. By clicking on specific areas, drop-down menus to switch combinations of variables, users can focus on critical questions in real time (e.g., "What is the critical state of the system at a certain parameter threshold?" and "What is the pattern of the spatial distribution of anomalous data points?"). Typical interaction strategies include the following:

- **Detail extraction:** This includes scaling from a macro-total (e.g., global climate patterns) to a micro-local (e.g., extreme precipitation events in each region).
- **Linked mining:** This includes interconnecting multiple datasets (e.g., brightening earthquake epicenter position in sync with underground rock structure data) to find cross-dimensional correlations.
- **Hypothesis testing:** By adjusting the parameters to simulate different scenarios (such as changing the proportion of new energy resources and observing the dynamic response of the grid load), it assists in theoretical deduction.

5. The cognitive layer: Sublimation from perception to insight

The goal is to achieve synergies at the above levels and convert a user's perception (e.g., "This color changes rapidly" and "That region looks like a whirlpool") into rational insights (e.g., "There is periodic oscillation in the physical field" and "The functional partition of the biological structure conforms to certain laws"). Digital media's creative language plays a key role in this process — dynamic effects direct attention to key changes, interactive actions reduce cognitive load, multimodal feedback enhances memory and understanding, and ultimately helps scientists discover patterns, anomalies, or new hypotheses in data faster.

2.4.3. *Paradigms of Integration: Innovative Scientific Visualization Across Disciplines*

The integration of scientific visualization and digital media has produced breakthrough applications in a few subject areas. The following three examples illustrate the logic of the integration:

Case 1: Dynamic visualization of cosmic evolution in astronomy
Study subjects: Galaxy formation and large-scale structure of the universe.

Traditional astronomical data (such as the position of hundreds of millions of galaxies in the SDSS survey and redshift) can only show the distribution of local regions, and it is difficult to reflect the overall structure of the universe. After incorporating digital media technology, the researchers developed a dynamic visualization system:

- **Data layer:** Integrate the redshift values of galaxies (mapped as distance/time), mass (mapped by point size), and orientation of motion (mapped in vector arrows).
- **Presentation layer:** The timeline animation shows the process of galaxy aggregation from the birth of the universe (redshift $z = 1100$) to the present — the uniform distribution of early matter, and then the gradual formation of fibrous structures and voids under the gravitational influence of dark matter.
- **Interaction layer:** Users can adjust the time progress bar (observe the structure of a particular epoch), click on a single galaxy to see its properties (such as mass and age), or switch to 3D mode (fly inside a galaxy cluster via a VR device).
- **Cognitive value:** This includes helping astronomers validate models of dark matter distribution and intuitively relay the evolutionary story of the universe from chaos to order to the public.

Case 2: Dynamic structure visualization of proteins in life sciences
Study subjects: Protein folding process and molecular interactions

The atomic-level structure data for proteins (obtained through X-ray crystallography or cryo-electrons) contain three-dimensional coordinates of thousands of atoms, and traditional static models only show folding states at a given moment. After incorporating digital media, the visualization system achieves the following:

- **Dynamic rendering:** This involves real-time demonstration of protein dynamics from linear peptide chains to functional folded states (e.g., hydrogen bond formation and secondary structure emergence), driven by molecular dynamics analog data.
- **Multimodal interactions:** Users can touch-spin 3D models to observe details, hover specific atoms to see chemical properties (such as

electronegativity and charge distribution), or play a sound to indicate key folding steps (such as the moment a hydrophobic core form).

- **Scientific significance:** Assist biologists in understanding the mechanisms by which proteins fail to function (such as diseases caused by misfolding) and provide dynamic references to target structures for drug design.

Case 3: Simulation visualization of extreme weather events in climate science

Study subjects: Formation and path prediction of typhoons/hurricanes

The grid data output by meteorological models (including air pressure, temperature, wind speed, and other parameters with a resolution of 1 km x 1 km) makes it difficult to visualize the complex structure of the storm through traditional wire graphs. After incorporating digital media, the visualization system integrates the following:

- **Dynamic rendering:** This simulates the rotation of the typhoon's eye and the movement of the spiral rain belt in real time and represents the wind speed strength through a color gradient (blue → red).
- **Interaction analysis:** Users can adjust simulated parameters (such as ocean surface temperature and atmospheric humidity) to observe the effects on the strength of the typhoon or switch to AR mode (scan map with tablet device and overlay virtual typhoon path and real terrain).
- **Application value:** This includes helping meteorologists optimize early warning models while intuitively communicating scientific explanations for why the path of this year's typhoon was abnormal to the public.

Future trends and core values of integration

The integration of scientific visualization and digital media is essentially a combination of scientific rigor and media creativity, revealing natural laws in a way that is more in line with human cognitive laws. In the future, with artificial intelligence (such as automated data labeling and intelligent recommendations to visualize perspectives), the convergence will become more intelligent, immersive, and pervasive, thanks to technologies such as augmented reality (XR) devices (such as lightweight AR glasses) and quantum computing (processing more complex data models). Its core value lies not only in improving the efficiency of scientific research (accelerating scientific discovery) but also in promoting scientific dissemination (making complex scientific phenomena understood by the public) and ultimately becoming a key bridge between data, scientists, and society. In the following chapters, we discuss the implementation details of this fusion logic in specific technical implementations, such as scientific visualization tool chain design and interdisciplinary collaboration processes.

Part 2

Technology: Digital Media Tool Chain and Scientific Visualization Methods

Chapter 3

Acquisition, Preprocessing, and Structuring of Scientific Data

3.1. Introduction: From "Raw Signals" to "Structured Knowledge" — The "Extraction-Purification-Modeling" Journey of Scientific Data

In the whole chain of scientific visualization, data are not a natural "ready-made material" but a precise transformation from the "physical world/ digital experiment" to the "visual canvas." It begins with the original signal (such as the starlight received by the telescope, sensor-recorded temperature fluctuations), measurement records of experimental processes (e.g., pH changes in chemical reactions and gene expression in biological samples), or outputs of computational simulations (e.g., grid data for fluid dynamics and probability distributions in quantum mechanics). But these raw data are often ungratified "pulp": They may be in mixed formats (such as text logs and binary files), noise-disruptive (such as anomalous values caused by sensor drift), dimensionally redundant (such as containing hundreds of variables but only a few related to the research goal), and, crucially, they usually exist in the form of "unstructured" or "semi-structured" (such as the lack of a unified timestamp format of time series data and spatial data without associated geographic coordinate system) and are difficult to be directly recognized and expressed by visualization tools (Kitchin, 2022; Ware, 2020).

The acquisition, processing, and structuring of scientific data is the "front-end engine" of this transformation — it is a complete process from "data acquisition source" to "visual input port." The goal is to enable raw data to undergo layers of extraction–purification–modeling, ultimately turning it into a structured dataset that is computable, interpretable, and visually friendly. This process is not only a technical operation (e.g., data cleansing and formatting) but also a representation of scientific thinking (e.g., clarifying the need for data dimensions for research objectives). Identifying the boundaries between key variables and noise is also a basic prerequisite for visual design (e.g., structured data can be accurately mapped into visual elements such as color, position, flow, and so on) (Sacha *et al.*, 2020).

3.2. Typical Types of Scientific Data

Scientific data are the cornerstones of scientific research, and their sources cover many fields such as natural sciences, engineering technology, and life sciences, carrying quantitative information on complex processes such as material movement, energy conversion, and life activities in the objective world. Unlike general commercial data or social media data, scientific data usually have the characteristics of high accuracy, strong domain dependence, and multi-dimensional coupling, and their collection methods, storage formats, and semantic connotations are closely related to the research goals of specific disciplines. Understanding the typical types and characteristics of scientific data is the prerequisite for subsequent data preprocessing and structuring, and it is also the basis for scientific visualization design. This chapter systematically analyzes the generation logic, structural characteristics, and application scenarios of various scientific data from the four core categories of observation data, experimental data, simulation data, and multimodal fusion data.

3.2.1. *Observational Data: Direct Records of Natural Phenomena*

Definition and Source: Observational data are the objective physical quantity, chemical quantity, or biomass information that exists in nature or social systems directly obtained through sensors, instruments, or manual recordings. This type of data is usually the result of "passive collection" — the researcher does not interfere with the state of the observed subject but only records its natural state through equipment or tools. Observational data are widely used in astronomy, meteorology, geology, ecology, environmental monitoring, and other fields and are the earliest and most basic data type in scientific research.

Typical Fields and Characteristics:

- **Astronomical observation data:** These include electromagnetic radiation signals obtained by optical telescopes (visible/infrared/ultraviolet bands), radio telescopes (radio bands), and space probes (e.g., Hubble Telescope and James Webb Telescope). Data forms include the spectra of galaxies (wavelength-intensity distribution), time series data of stars' brightness over time (light curves), and surface temperature distributions of planets (infrared imaging). The characteristics are high-dimensional (multi-band parameter coupling), large-scale (covering the space-time range of the universe), and sparsity (some celestial phenomena only occur occasionally). For example, the LSST (Large Integrated Survey Telescope) will generate more than 20 terabytes of galaxy image data every night, and it needs to track transient events such as supernova explosions through time series analysis (LSST Science Collaboration, 2023).
- **Meteorological observation data:** These include atmospheric state parameters collected by ground weather stations (temperature, humidity, air pressure, and wind speed), sounding balloons (vertical atmospheric profile), satellite remote sensing (cloud cover and sea surface temperature), radar (precipitation particle distribution), and other equipment. Data are stored in the form of grids (e.g., global latitude and longitude grids) or station-based (discrete observation points), with temporal resolutions ranging from minutes (e.g., minute rainfall) to hourly (e.g., conventional meteorological elements), and spatial resolutions ranging from 100 meters (urban weather stations) to kilometers (satellite remote sensing). For example, typhoon track prediction needs to be corrected in real time by integrating multi-source observation data (satellite cloud image + radar echo + ground wind speed) (World Meteorological Organization (WMO), 2021).

- **Ecological environment observation data:** These include ecosystem parameters obtained through UAV remote sensing (vegetation coverage), ground sensor network (soil pH and microbial activity), satellite multi-spectral imaging (chlorophyll content and forest biomass), and other means. Data often exist in the form of multispectral/hyperspectral imagery (hundreds of narrow-band reflectance values) or point sampling data (e.g., stratified measurements of soil profiles) and are characterized by multivariate coupling (biological–physical–chemical factor interactions) and long-term (need to be monitored for many years to identify trends). For example, in the study of the global carbon cycle, it is necessary to integrate multi-source observation information such as forest canopy height (LiDAR data), soil organic carbon content (borehole sampling), and atmospheric CO_2 concentration (time series data from monitoring stations) (Clark *et al.*, 2022).

3.2.2. *Experimental Data: Evidence of Variable Association Under Controlled Conditions*

Definition and Source: Experimental data are quantitative information obtained by actively intervening (such as changing parameters such as temperature, pressure, and chemical reagent concentration) and measuring the response of the system in a human-controlled experimental environment. The core feature of this type of data is the "control-observation" logic — through the researcher design control experiments (such as setting up experimental and control groups) and control variables (keeping other conditions unchanged and adjusting only a single factor) to clarify the causal relationship or correlation between independent variables and dependent variables. Experimental data are the core research material of basic disciplines such as physics, chemistry, biology, and materials science and are also the key basis for verifying theoretical hypotheses in engineering technology (such as new drug research and development, and chip process optimization).

Typical Fields and Characteristics:

- **Physical experiment data:** These include microscopic particle behavior or macro-physical phenomenon parameters obtained by particle accelerators (such as the Large Hadron Collider, LHC), superconducting quantum interferometer (SQUID), vacuum chamber experimental equipment, etc. For example, the energy, momentum, and flight trajectory of secondary particles generated by proton collisions in LHC experiments (measured by electromagnetic energy detectors and track detectors) are stored in the form of high-dimensional event records (each collision event contains multi-parameter information of hundreds of particles), characterized by ultra-high accuracy (measurement error of less than 10^{-12}) and event sparsity (the proportion of target signal events is extremely low, and it needs to be filtered from massive background noise) (CERN, 2021).

- **Chemical experiment data:** These include molecular structure, reaction rate, or substance composition information measured by instruments such as spectrometers (infrared/NMR/mass spectrometry), chromatographs (gas/liquid), and electrochemical workstations. Typical data include the molecular fingerprint of the compound (e.g., absorption peak position and intensity in the infrared spectrum), the pH curve of the reaction system over time, and the elemental distribution on the surface of the catalyst (electron microscopy image). The characteristics are multimodal coupling (structure–property–reaction path correlation) and small sample (the experimental cost of some complex reactions is high, and the data amount is limited). For example, in the development of new drugs, tens of thousands of compounds need to be screened through high-throughput experiments, and their binding affinity (IC50 value) and toxicity parameters to target proteins are recorded (Smith *et al.*, 2020).
- **Biological experimental data:** These include experimental measurements covering molecular biology (e.g., gene expression), cell biology (e.g., cell proliferation rate), physiology (e.g., ECG signal), etc. For example, gene transcript abundance data obtained by RNA sequencing technology (RNA-seq) (expressed as TPM/FPKM values), cell migration trajectories taken by microscopy (x–y–t 3D coordinate sequence), and blood pressure/heart rate monitoring timing data from animal experiments. It is characterized by high dimensionality (a single sample may contain tens of thousands of gene expression values), heterogeneity (significant differences in data formats and units generated by different experimental platforms), and biological noise (random errors introduced by individual differences or environmental fluctuations) (ENCODE Project Consortium, 2022).

3.2.3. *Simulation Data: Digital Mapping of Theoretical Models*

Definition and Source: Simulation data are the data generated by digital deduction of real physical processes or complex system behavior based on mathematical models (such as partial differential equations, statistical mechanics models, and system dynamics models) through computer numerical simulation technology. Simulation becomes a central means to study these problems when actual observations or experiments are limited by cost, ethical, or technical conditions (such as the inability to directly observe the flow of the Earth's deep mantle and the inability to repeat the early state of the Big Bang). Simulation data are the product of "theoretical assumptions, mathematical modeling, and numerical solving," and their accuracy depends on the rationality of the model and the accuracy of the calculation method.

Typical Fields and Characteristics:

- **Engineering simulation data:** These include parameter distribution generated by fluid dynamics simulation (CFD, computational fluid dynamics), structural mechanics simulation (finite element analysis, FEA), electronic circuit simulation (SPICE model), etc. in mechanical engineering. For example, in the aerodynamic shape design of an aircraft wing, CFD simulation outputs the surface pressure distribution (Pascal value), airflow velocity vector (m/s), and eddy current area (identified by streamline visualization). In the seismic design of buildings, FEA simulation calculates the stress–strain (Newton/m^2) of beam-column joints under different seismic wave inputs. The characteristics are high spatial resolution (grid size up to millimeters) and parameter coupling (multi-field interaction of velocity field, pressure field, and temperature field) (ANSYS, 2023).
- **Simulation data of scientific theory:** These include spatial–temporal dynamic data generated by quantum many-body system simulations in physics (such as electron correlation effects of superconducting materials), galaxy evolution models in astrophysics (aggregation of matter under the gravitational action of dark matter), and global circulation models (GCM, coupled atmosphere-ocean-land surface processes) in climate science. For example, climate models. The forecast value of temperature, precipitation, and sea level rise in the next 50 years (time step from hours to months, spatial step size is 1 × 1

km) of the global grid per kilometer and the data volume can reach petabyte level. In quantum chemistry simulations, the distribution of molecular orbital energy levels (in electron volts) and the transition state energies (activation energy) of reaction pathways are key features. It is characterized by high dimension (multi-physics + long-term scale), uncertainty (error transfer of model parameters), and computational intensity (a single simulation may require several weeks of supercomputer operation) (IPCC, 2021; Tuckerman *et al.*, 2022).

3.2.4. *Multimodal Fusion Data: Collaborative Expression of Cross-Source Information*

Definition and Source: Multimodal fusion data refer to the integration of heterogeneous data sources from different sensors, observation methods, or theoretical models (such as text description + image + numerical parameters, structural data + temporal data + spatial data) to more comprehensively describe the attributes and behaviors of complex scientific objects. The core value of this type of data lies in solving the problem of missing or ambiguous information from a single data source through the complementarity and confirmation of multi-dimensional information, which is commonly found in cutting-edge fields such as precision medicine, earth system science, and intelligent robots.

Typical combinations and characteristics:

- **Medical multimodal data:** These include imaging data (structural images of CT/MRI/X-rays), physiological signals (timing waveforms of ECG and EEG), omics data (genomic DNA sequences and proteome mass spectrometry peaks), and clinical texts (symptom descriptions in medical record reports) integrated during patient diagnosis and treatment. For example, tumor diagnosis needs to integrate tumor morphology (spatial location and boundary) from MRI, metabolic activity (glucose uptake rate) from PET scans, cell grading from pathological sections (microscopic images), and genetic mutation information from patients (NGS sequencing data) to formulate personalized treatment plans. It is characterized by large-scale differences between modalities (from molecular-level genetic data to organ-level images) and complex semantic associations (different modalities reflect various aspects of the same pathological process).

- **Earth system multimodal data:** These include satellite remote sensing data (cloud optical thickness and sea surface temperature), ground observatory data (soil moisture, precipitation), reanalysis data (interpolation of global meteorological models), and socio-economic data (energy consumption and land use type) integrated in climate studies. For example, drought monitoring needs to integrate the number of consecutive rainless days (time series data) of the weather station, the vegetation index (NDVI) inverted by the satellite, and the deep moisture content of the soil moisture sensor network (spatial grid data) to assess the spatiotemporal scope and impact of drought. It is characterized by inconsistency in spatiotemporal resolution (satellite data are global coarse resolution but updated at high frequency, and ground data are local high resolution but low frequency) and interdisciplinary semantics (physical parameters and social behavior need to be modeled uniformly) (NASA Earthdata, 2022).

The Need for and Challenges of Integrating Different Types of Scientific Data:

The above four types of typical scientific data (observation, experiment, simulation, and multimodal fusion) together constitute the "data panorama" of scientific research, and their core commonality lies in the coexistence of high value and complexity — they not only are key evidence to reveal the laws of nature but also put forward higher requirements for subsequent preprocessing and structuring due to problems such as format heterogeneity, dimensional coupling, and noise interference. For example, spatiotemporal alignment of observation data (such as registering remote sensing images from different satellites to a unified coordinate system), variable screening of experimental data (identifying key influencing factors from hundreds of parameters), grid dimensionality reduction of simulation data (reducing computational and storage costs), and feature alignment of multimodal data (mapping image pixels and gene sequences to a unified analysis space) are all issues that need to be addressed in the following chapters (preprocessing and structuring). Understanding the characteristics of these typical data types is the starting point of the entire process management of scientific data, and it is also the basis for scientific visualization design to "adapt to local conditions."

3.3. Data Sources and Formats

The value of scientific data is not only dependent on its content itself but also closely related to the characteristics of its source channels and the specifications of storage formats. Scientific data from different sources often carry unique meta-information (such as acquisition equipment parameters, experimental environmental conditions, and model assumptions), and format differences directly affect the readability, compatibility, and processing efficiency of the data. This chapter systematically explains the "ins and outs" and "storage logic" of scientific data from two aspects: the classification and analysis of main source channels and the technical characteristics of common storage formats and lay the foundation for subsequent preprocessing and structuring operations.

3.3.1. *Data Source: Multi-Subject and Multi-Scenario Acquisition Network*

The sources of scientific data can be summarized into four core channels, each corresponding to different collection subjects, technical means, and application scenarios, and their data characteristics and quality assurance mechanisms are also significantly different.

1. Research Institutions and Laboratories (Professional and Controllable Sources)

Subjects and scenarios: Universities, national research institutes (such as the Chinese Academy of Sciences and US national laboratories), and enterprise R&D centers (such as drug laboratories of biopharmaceutical companies) are the core producers of scientific data. These institutions carry out observations, experiments, or simulation studies in controlled environments (such as constant temperature laboratories, vacuum chambers, and clean workshops) or field fixed-point (such as weather stations and ecological observation stations) by independently designing or purchasing professional equipment.

Typical data and characteristics:

- **Experimental data:** These include high-precision parameters generated by mass spectrometers (measuring molecular mass-charge ratios) in chemistry labs, fluorescence microscopy (observing protein localization in cells) in biology labs, and superconducting quantum interferometers in physics labs (detecting weak magnetic field changes), usually accompanied by detailed experimental conditions (e.g., temperature, pH, and reagent concentration) and operation logs (e.g., sample preparation steps and instrument calibration records).

- **Observation data:** The original signals collected by the observatory's telescope array (such as Guo Shoujing's telescope LAMOST), the seismic network of the Institute of Geology (recording crustal vibration waveforms), and the CTD profiler of the oceanographic research vessel (measuring sea temperature, salinity, and depth) are often stored in the form of "original electrical signal + metadata (equipment model, sampling frequency)," which needs to be preliminarily converted by professional software (such as IRAF astronomical image processing package and SEISAN seismic analysis system).
- **Simulation data:** These include simulation results generated by research teams based on self-developed numerical models (such as regional versions of climate models and molecular dynamics codes for new material designs) or open-source tools (such as OpenFOAM fluid simulation software and LAMMPS molecular simulation package), usually include model parameters (such as mesh size and boundary conditions), computational configuration (such as number of parallel nodes and number of iterations), and convergence verification information.

Quality assurance: These data generally follow strict experimental design specifications (e.g., randomized controlled trials and repeated measurements) or model validation processes (e.g., benchmark cases), have a high degree of metadata integrity (record device serial number and software version number), but may have "data silos" (formatting standards vary from lab to lab).

2. Government and Public Sector (Macro Monitoring Sources)
Subject and scenario: Meteorological bureaus (such as China Meteorological Administration, CMA, and US NOAA), environmental protection departments (such as the Ministry of Ecology and Environment), earthquake bureaus (such as China Earthquake Administration), and space agencies (such as NASA and ESA), etc. continue to collect public science

data covering a wide range and long-term scale by deploying national monitoring networks or launching space exploration equipment.

Typical data and characteristics:

- **Environmental monitoring data:** These include grid or point data generated by air quality monitoring stations (PM2.5 and SO_2 concentration), automatic water quality monitoring buoys (COD and heavy metal content), and soil pollution survey projects (full and effective state of heavy metals) of environmental protection departments, usually indexed by "administrative division + timestamp," and the data update frequency varies from hourly (air quality) to daily (water quality), with monitoring method standards (such as national standard GB number) and quality control marks (such as whether it has passed laboratory comparison).
- **Earth observation data:** These include global surface images (reflectivity and temperature), sea surface temperature (SST), and vegetation index (NDVI) obtained by multispectral/hyperspectral sensors carried by meteorological satellites (such as Fengyun series) and earth observation satellites (such as Landsat 8 and Sentinel-2), and the data are stored in "Scene (fixed coverage)" storage, including orbital parameters (satellite altitude and inclination), imaging time (UTC time), radiation calibration coefficient (conversion of the original DN value into physical quantity), and other meta information.
- **Space exploration data:** These include planetary surface composition, magnetic field strength, and atmospheric composition data transmitted by spacecraft payload instruments (such as the rover's laser-induced breakdown spectroscopy, LIBS, and Jupiter rover's magnetometer), usually accompanied by the probe's position and attitude information (latitude, longitude, altitude, and yaw angle) and mission stage markers (such as before/after landing).

Quality assurance: Such data must comply with national or international standards (such as WMO Meteorological Data Specification and ISO Geographic Information Standard) and be published through official channels (such as NOAA's data portal and NASA's Earth data platform), but there may be limitations in spatiotemporal resolution (such as satellite revisit periods of several days) or uneven regional coverage (such as sparse data in polar regions).

3. Business Platforms and Sensor Networks (Market-Based Supplementary Sources)

Subjects and scenarios: Commercial satellite companies (such as Maxar and Planet Labs), IoT companies (such as Alibaba Cloud IoT and HUAWEI CLOUD IoT), and crowdsourcing science platforms (such as Zooniverse citizen science projects) provide supplementary data sources for scientific research by deploying low-cost, high-density sensor networks or providing commercial data services.

Typical data and characteristics:

- **Commercial remote sensing data:** Sub-meter resolution optical imagery (0.5–1 meter pixels) provided by tiny satellite constellations (such as Planet Labs' "pigeon flock" satellites) can be updated multiple times a day, covering any region of the world, usually in GeoTIFF (geotagged raster image) or custom binary files, with auxiliary parameters such as cloud cover ratio and solar altitude angle.
- **IoT monitoring data:** Real-time timing data generated by soil moisture sensors in the agricultural field (monitoring 0–100 cm soil moisture), equipment vibration sensors in industrial scenarios (predicting mechanical failures), and environmental noise monitors in smart cities (decibel values and spectrum distribution) are transmitted to cloud platforms through low-power networks, such as LoRa/NB-IoT.
- **Crowd sourced observation data:** These include the distribution location of birds (GPS coordinates + time) recorded by citizen scientists through mobile apps, meteor shower images taken by astronomers (with timestamps and exposure parameters), and the subjective perception of air quality uploaded by community residents (such as odor occurrence time), and the data have the characteristics of "long-tail distribution" (a small amount of high-quality data and a large number of low-precision data coexist), which need to be filtered by quality control algorithms (such as denoising and consistency check).

Quality assurance: The quality of these data varies (depending on equipment cost and level of maintenance) and metadata may be incomplete (e.g., missing sensor calibration records), but it is flexible (quick response to specific research needs) and wide coverage (e.g., crowd-sourced sighting reports of forest fires around the world).

4. Open-Source Community and Sharing Platform (Collaborative Sharing Source)

Subjects and scenarios: Academic open-source communities (such as GitHuband GitLab), scientific data sharing platforms (such as Figshare, Zenodo, and Kaggle's scientific dataset section), and interdisciplinary cooperation projects (such as the Human Genome Project and Global Earthquake Alliance) promote the reuse and collaborative analysis of scientific data through open data policies (Wilkinson, *et al.*, 2016).

Typical data and characteristics:

- **Open-source scientific research datasets:** The application of machine learning in science (e.g., the AlphaFold database for protein structure prediction), social science surveys (e.g., the World Values Survey's cross-border questionnaire data from WVS), and astronomical catalogs (e.g., the position and motion parameters of a billion stars on the Gaia satellite) are published in structured tables (CSV/JSON) or databases (SQL/NoSQL), usually with data dictionaries (field meaning descriptions) and license agreements (e.g., CC-BY 4.0).
- **Collaborative project data:** Massive experimental data (petabytes) generated by large international scientific installations (such as CERN's Large Hadron Collider LHC) are shared with members of the global collaboration group through distributed storage systems (such as EOS) and dedicated data formats (such as ROOT format and high-energy physical data containers developed by CERN), and data access needs to be controlled (based on the contribution of the research team).

Quality assurance: Such data typically undergoes initial cleaning and standardization (e.g., in accordance with the FAIR principles: Findable, Accessible, Interoperable, and Reusable), but compatibility issues may arise due to version iterations (e.g., updates to open-source projects), so pay attention to the version number and changelog in the metadata.

3.3.2. *Data Format: Coding Logic from the Original Signal to the Structured Record*

The format of scientific data is determined by the technical requirements of its source channel and can be divided into three categories: original format (direct equipment out), intermediate format (domain general), and standard format (cross-domain compatibility). Each format has its own specific coding rules and application scenarios.

1. Raw Format: "First-Hand Data" Directly From the Device

Definition and Characteristics: Raw format is an unprocessed binary or text file generated directly by a data acquisition device, usually containing the lowest level of electrical signals, pixel values, or sensor readings, preserving the most complete original information but lacking readability.

Typical examples:

- **Image class:** RAW format for scientific cameras (e.g., TIFF 16-bit uncompressed images) that stores the original photodiode response value per pixel (without white balance, gamma correction) contains full dynamic range (e.g., 14-bit depth recording 0–16383 grayscale). The FITS format (Flexible Image Transport System, developed by the International Astronomical Union IAU) of the astronomical CCD embeds metadata such as observation time, telescope pointing, filter wavelength, and other metadata (stored in the form of header keywords) in addition to image pixels.
- **Timing class:** The electrical signals collected by the oscilloscope (such as voltage–time curves) are stored in binary files, and the numerical values (such as 16-bit integers) and timestamps (nanosecond accuracy) of each sampling point are recorded. The SEED format (Standard for the Exchange of Earthquake Data) of the seismograph contains continuous sampling data of seismic waveforms and station metadata (such as instrument response functions).
- **Experimental:** The raw data file generated by the mass spectrometer (e.g., Thermo Fisher's .raw format) stores the time-of-flight, mass-to-charge ratio (m/z), and intensity values of ions, along with instrument parameters (e.g., ion source temperature and collision energy). The gene sequencer's FASTQ format (text format but raw sequencing signal) records the sequencing mass fraction (Phred value) and raw fluorescence signal intensity for each base.

Processing requirements: The original format needs to be converted to an intermediate format that can be analyzed by professional software (such as ImageJ processing FITS images and MATLAB reading binary timing files) or instrument supporting tools; otherwise, it is difficult to interpret directly.

2. Intermediate format: "Structured records" for domain use

Definition and characteristics: Intermediate format is a semi-structured format widely used in specific disciplines or technical fields, usually for preliminary collation of raw data (e.g., noise removal and timestamp alignment), and storage of key parameters in a more readable text or binary structure, taking into account readability and computational efficiency.

Typical examples:

- **Text class:** CSV (Comma-Separated Values) is the most common intermediate format for storing tabular data (e.g., temperature–time curves measured by experiments, each row represents a point in time, columns are time, temperature, and humidity); JSON (JavaScript Object Notation) stores nested data (such as the configuration parameters of the sensor network, including device ID, sampling frequency, and list of measurement variables) in the form of key-value pairs, which is suitable for describing non-regular structures.
- **Binary:** HDF5 (Hierarchical Data Format version 5), developed by the National Supercomputing Application Center (NCSA), supports the storage of multi-dimensional arrays (such as three-dimensional temperature fields of climate models), metadata (such as variable units and coordinate system definitions), and hierarchical structures (such as sub-datasets organized by year/region), which are widely used in earth science, high-energy physics, and other fields. NetCDF (Network Common Data Form) focuses on the self-descriptiveness of scientific data (annotating variable meanings through attributes) and is commonly used in meteorology (e.g., global wind field data stored by NetCDF4) and oceanography (e.g., SSE-salinity profiles).
- **Field-specific category:** This includes DICOM (Digital Imaging and Communications in Medicine), standard format for medical imaging (CT/MRI/X-ray), including pixel data, patient information (name and age), scanning parameters (layer thickness and voltage), and image geometry information (pixel pitch, orientation matrix); ROOT (developed by CERN), high-energy physics experiments. The container format supports efficient storage of billions of particle event records

(each event contains the momentum, energy, and position of multiple particles) and enables fast retrieval through the C++/Python interface.

Advantages and limitations: Intermediate formats balance data structuring and tool compatibility (e.g., Python's pandas' library can directly read CSV and h5py library can operate HDF5), but intermediate formats in different fields may have syntax differences (e.g., NetCDF and HDF5 have different metadata annotation rules), which need to be dealt with in a targeted manner.

3. Standard format: A "universal language" that is compatible across domains

Definition and characteristics: Standard format is a standardized format developed to facilitate multidisciplinary data sharing and interoperability, typically published by international organizations (e.g., ISO and W3C) or industry alliances (e.g., OGC, Open Geospatial Information Consortium), emphasizing metadata integrity and cross-platform compatibility.

Typical examples:

- **Geospatial data:** GeoJSON (JSON-based geographic data format), which describes spatial elements such as points, lines, and surfaces (such as the coordinate set of urban boundary polygons) and attributes (such as population density) in a standardized syntax, compatible with web map services (such as Leaflet.js front-end display); GML (Geography Markup Language, XML-based), an XML format formulated by OGC, support topological relationships for complex geographic objects (e.g., water system connectivity for river networks).
- **Common standards for scientific data:** These include CF (Climate and Forecast) metadata conventions, NetCDF files for climate and meteorological data, mandatory annotation of the physical meaning (in K or °C), coordinate system (latitude and longitude grid or projected coordinate system), time reference (UTC timestamp), and grid topology (e.g., longitude/latitude resolution) to ensure data comparability of global climate models; ISO 19115 (Geographic Information Metadata Standard), which defines comprehensive description information of spatial datasets (including data producers, collection methods, accuracy assessments), applies to public science data released by governments.

- **Multimodal fusion standard:** These include DICOM-SR (DICOM Structured Reporting), which extends the structured coding of diagnostic reports (such as tumor staging results) on the basis of medical images; FHIR (Fast Healthcare Interoperability Resources), which enables cross-system sharing of electronic medical records, laboratory test results, and imaging data through RESTful APIs.

Value and challenge: Standard formats are the cornerstone of long-term scientific data preservation and global collaboration (e.g., NetCDF data for CF standards for IPCC climate change reports), but compliance requires additional metadata annotation work (e.g., adding CF-compliant attributes to each variable), and data generated by some older devices may require format conversion tools (e.g., converting legacy DICOM to DICOM-SR).

Summary: The synergistic impact of source and format
The source of scientific data determines its initial form (e.g., raw signals from the device or structured parameters from model output) and quality characteristics (e.g., precise control of laboratory data vs. uncertainty of crowdsourced data), while format constrains data storage efficiency (e.g., binary format saves space but cannot be read directly), readability (e.g., text format is easy to understand but highly redundant), and compatibility (e.g., standard format supports cross-platform but requires additional annotation). In actual scientific research, data usually need to go through the process of "original collection → intermediate collation → standard conversion" (e.g., the original DN value of satellite remote sensing is first converted into physical quantities after radiation calibration and then stored as NetCDF files according to CF standards), which is the core task of "preprocessing" in the subsequent chapters. Understanding the diversity of data sources and formats is the first step in the full lifecycle management of scientific data, and it is also the prerequisite for efficient visualization and analysis (Sacha *et al.*, 2020).

3.4. Data Cleaning and Preprocessing

Scientific data often goes through a series of data cleaning and preprocessing operations from raw acquisition to a "clean, usable" state that can be used for visualization or analysis. This is the most critical and time-consuming part of the scientific data processing process — raw data are often noisy, missing, redundant, or inconsistent due to collection errors, equipment limitations, environmental interference, or formatting irregularities, which can lead to incorrect pattern recognition (e.g., misjudging noise as a trend), misleading conclusions (e.g., distorting statistical results due to improper filling of missing values), or inefficient calculations (e.g., redundant data taking up too much storage and computing power). This chapter systematically analyzes how to transform "raw data" into "high-quality scientific data assets" from the aspects of the core objectives of cleaning, typical problems and solutions, as well as the key steps and technical logic of preprocessing.

3.4.1. *Data Cleaning: Fix the Quality Cornerstone of "Dirty Data"*

The core goal of data cleaning is to identify and correct errors, inconsistencies, or outliers in the data, ensuring its accuracy, integrity, and consistency. Due to the specificity of scientific data, common "dirty data" problems can be divided into the following four categories, each of which has a corresponding cleaning strategy.

1. Missing Value Handling: The Art of Trade-Offs for Filling or Eliminating

Problem manifestations: Observation equipment failure (such as temporary sensor failure of weather station), experimental interruption (such as missing part measurements due to biological sample contamination), and transmission errors (such as loss of records due to network packet loss) can lead to the loss of certain fields or samples in the dataset. Missing modes may be completely random missing (MCAR, such as randomly broken sensors), random missing (MAR, such as missing data at only some stations due to weather in a certain period of time), and non-random missing (MNAR, such as the instrument automatically stopping recording when critical parameters exceed the standard); the latter two should be handled carefully to avoid bias.

Cleaning strategy:

- **Listwise deletion:** This directly deletes the entire record (e.g., all sensor data at a point in time) or the entire field (e.g., all measurements of a faulty instrument) containing missing values. It is suitable when the proportion of missing values is extremely low (<5%) and the missingness is MCAR (e.g., minute-level temperature data from a specific weather station is missing; deletion of this data does not affect the overall trend).

- **Imputation:** This is the estimation of missing values through statistical or model methods. Common techniques include the following:
  - **Statistical imputation:** Fill with mean (continuous variable), mode (categorical variable), or median (anti-outlier) (e.g., fill in the missing value of a single day with the average temperature of a station in the past week).
  - **Proximity interpolation:** This is based on spatiotemporal proximity (e.g., interpolation with synchronous data from adjacent weather stations and data from the same sensor before and after time points, suitable for scenarios with strong spatial/temporal correlation).
  - **Model prediction:** Predict missing values (e.g., predicting missing PM2.5 concentrations based on other environmental parameters (humidity and wind speed)) by regression (continuous variables), classification (categorical variables), or machine learning models (e.g., random forests and KNN).
- **Flagging:** Keep the missing value but add a binary tag field (e.g., "is missing = 1") to explicitly consider the impact of the missing in subsequent analysis (e.g., missing as a feature input in a machine learning model).

2. Noise and Outlier Detection: Distinguish between real signals and interference

Problem manifestation: Random fluctuations in observation equipment (such as minor voltage jumps caused by electronic noise), sudden environmental interference (such as the instantaneous impact of lightning on meteorological sensors), and human recording errors (such as erroneous input of extreme values) can generate noise (high-frequency and small-scale fluctuations) or anomalies (outliers that significantly deviate from the normal range). Anomalies in scientific data can either represent genuine rare events (such as the instantaneous high-energy radiation from a supernova explosion or extremely elevated indicators of a rare disease) or erroneous data, which requires the integration of domain-specific knowledge for accurate interpretation.

Cleaning strategy:

- **Statistical methods:** Based on the normal distribution assumption (applicable to approximate normal data), outliers are identified by the using the 3σ principle (values exceeding $\mu \pm 3\sigma$ are considered anomalous if the data follows $N(\mu, \sigma^2)$), or box plot method (IQR = interquartile range, values above Q3 + 1.5 × IQR or below Q1 − 1.5 × IQR are abnormal). For non-normal distributions (e.g., skewed biological indicators), it can be adopted to recalculate the threshold with logarithmic transformation.

- **Field rules:** These set a reasonable range according to scientific common sense (e.g., the normal range of human body temperature is 36–37.5°C, and the measured value beyond this range needs to be verified; in meteorological data, if the instantaneous wind speed in a certain place exceeds 2 times the historical extreme value, it may be an instrument failure).
- **Smoothing filtering:** This reduces the effect of random fluctuations on time series data (such as physiological signals continuously monitored by sensors) through moving averages (MA, averaging values close to N points), Savitzky–Golay filtering (smoothing high-frequency noise while preserving trends), or wavelet transform (separating the low-frequency component of the signal from high-frequency noise).
- **Visual assistance:** This visually locates outliers (such as sudden spikes in line charts) through box plots, scatter plots, or time series plots and determines whether they are real events (such as abnormal global temperature increases due to volcanic eruptions) based on context.

3. Duplicate and redundant data: Eliminate the burden of invalid information

Performance of the problem: This is a multi-source data combination due to inconsistent field names (such as "temperature" and "TEMP" being the same variable). The collection equipment repeats records (e.g., sensors sampling every second but only minute-level data), and the misleading entry of backup files results in identical records or multiple field repeated information in the dataset. Redundant data not only take up storage space but can also distort statistical results (e.g., repeated samples resulting in a high mean).

Cleaning strategy:

- **Deduplication operation:** This deletes exactly the same record based on a unique identifier (e.g., timestamp + sensor ID and sample number) (e.g., drop duplicates() function of the pandas library). For duplicate fields (such as "Place Name" in multiple tables but other parameters are different), you need to retain the key fields after merging them through primary key associations.

- **Down-sampling:** For high-frequency sampling data (e.g., vibration signals 10 times per second), this reduces the sampling frequency as needed (e.g., convert to the average value 1 time per minute) to retain the trend while reducing the amount of data (suitable for scenarios where microscopic details are not required in visualization).
- **Field merging:** This unifies synonymous fields from different sources (e.g., map "TEMP," "temperature," and "T" to "temperature") and remove redundant columns that don't make sense (e.g., "Data Entry ID" for internal records only).

4. Inconsistencies and formatting errors: Unify the semantics and structure of data

Problem manifestations: These include inconsistent units (e.g., temperature using °C and K at the same time), chaotic time format (e.g., "2025-09-06," "06/09/2025," and "September 6, 2025"), contradictory classification labels (e.g., "male" and "male" are considered different categories), and value range crossing the boundary (e.g., age is negative or exceeding the human limit). These issues can lead to logical errors in subsequent analyses (e.g., unit confusion leading to biased calculation results).

Cleaning strategy:

- **Unit normalization:** This converts all data to uniform units (e.g., temperature to °C, length to meters), converts by multiplier factor (e.g., K→°C: minus 273.15) or conversion formula, and indicates the original unit in the metadata.
- **Time alignment:** This converts time fields to standard formats (e.g., ISO 8601: YYYY-MM-DD HH:MM: SS) and unifies time zones (e.g., all to UTC) or time granularity (e.g., aggregate "2025-09-06 10:05:30" into "2025-09-06 10:00:00" time period).
- **Classification encoding:** This converts text-based categorical variables into numerical encoding (e.g., "male" = 1, "female" = 0), or unified text representations (e.g., "male" and "male" are mapped to "M"), and records the correspondence between the original and the encoding through a dictionary mapping table.
- **Range check:** This sets a reasonable value range based on domain knowledge (e.g., pH should be between 0 and 14 and heart rate should

be 30–250 beats/min), and values outside the range are marked as abnormal or corrected (e.g., negative age is corrected to missing values).

3.4.2. *Data Preprocessing: The Transformation Logic that Paves the Way for Visualization and Analysis*

Data preprocessing is based on the cleaned "clean data," through further transformation and construction, to make it more suitable for the visualization requirements or the input requirements of the analysis model. Its core goal is to improve data availability, expressiveness, and computational efficiency, mainly including the following key steps:

1. Data standardization and normalization: Eliminate the impact of dimensional comparisons

Background: The physical dimensions of different variables in scientific data vary greatly (e.g., the numerical range of temperature (°C) and pressure (Pa) differs by many orders of magnitude), used directly for visualization (e.g., scatter plots) or machine learning models (e.g., KNN and neural networks) results in quantitatively large variables dominating results (e.g., high values of pressure values masking small changes in temperature).

Pretreatment method:

- **Standardization (Z-score):** This converts the data into a distribution with a mean of 0 and a standard deviation of 1, with the formula $X_{\text{std}} = (X - \mu) / \sigma$ (μ is the mean, σ is the standard deviation), which is suitable for scenarios where the data are approximately normally distributed (such as most physical measurement parameters).
- **Normalization (Min–Max):** This scals the data linearly to a range of [0,1] with the formula $X_{\text{norm}} = (X - X_{\text{min}}) / (X_{\text{max}} - X_{\text{min}})$, suitable for scenarios with a clear data range and no extreme outliers (e.g., image pixel values (0–255) and questionnaire scores (1–5 points)).
- **Robust scaling:** This uses median and interquartile range (IQR) substitutions for means and standard deviations ($X_{\text{robust}} = (X - \text{median}) / IQR$) to combat outlier interference (e.g., economic data with a small sample of very high income).

2. Dimensionality reduction and feature construction: Simplifying complex information purification

Background: High-dimensional scientific data (such as multivariable sensor networks simultaneously monitoring dozens of parameters such as temperature, humidity, air pressure, and wind speed) can lead to "dimensional disasters" (crowded graphics during visualization and high computational complexity during analysis) or redundant variables (such as multiple sensors measuring different components of the same physical quantity).

Pretreatment method:

- **Dimensionality reduction technology:**
  - **Principal component analysis (PCA):** This converts the original variables into a set of unrelated principal components (sorted by variance contribution rate) through linear transformation and retains the first few principal components to explain most of the data variation (e.g., reducing 10 meteorological parameters to 3 principal components, covering more than 90% of the information).
  - **t-SNE/UMAP:** This is a nonlinear dimensionality reduction method, suitable for visualization of high-dimensional data (e.g., reducing gene expression data to 2 dimensions, showing clustering patterns of cell types) but more focused on the preservation of local structure rather than global variance interpretation.
- **Feature construction:** This generates new and more explanatory variables based on domain knowledge (e.g., "day-night temperature difference = daily maximum temperature — daily minimum temperature," "wind chill index = temperature × wind speed function") or combines original variables into composite indicators (e.g., air quality index AQI is calculated by weighting multiple pollutant concentrations such as PM2.5 and SO_2).

3. Spatial–temporal alignment and gridding: A spatial datum for a unified observation scale

Background: Multi-source scientific data often cannot be directly compared or fused due to differences in the location of the acquisition equipment (such as the longitude and latitude coordinates of different weather stations), inconsistent time steps (such as the transit time of satellites every hour and every 10 minutes for ground observation), or different spatial resolutions (such as the 1km grid of satellite remote sensing and the 10m resolution of UAV aerial photography).

Pretreatment method:

- **Space-time alignment:**
  - **Time alignment:** This includes aggregation of data with different time resolutions (e.g., averaging 10 minutes of temperature data to an hourly average) or interpolation (e.g., scaling hourly data to minutes by linear interpolation).
  - **Spatial alignment:** This converts data from different coordinate systems to a unified frame of reference (such as WGS84 latitude and longitude or UTM projected coordinates) and achieves spatial matching through resampling (e.g., down-sampling high-resolution imagery to the same grid size as low-resolution data) or interpolation (e.g., Kriging interpolation populates discrete weather station data into a continuous grid).
- **Meshing:** This converts discrete point observations, such as single-point measurements from a sensor network, into regular grids (such as latitude and longitude grids or custom area grids) for subsequent visualization rendering (e.g., heat mapping of temperature distributions) or spatial analysis (e.g., calculating averages within a grid).

4. Multimodal data fusion: Collaborative coding of cross-source information

Background: Heterogeneous data sources (such as satellite images + ground sensor data, gene sequences + clinical indicators) are often integrated in scientific visualization, but the data formats, scales, and semantics of different modalities are significantly different (e.g., the pixel matrix of the image cannot be directly correlated with the readings of numerical sensors).

Pretreatment method:

- **Feature-level fusion:** This extracts data from different modalities into unified feature vectors (such as extracting deep features from images through convolutional neural networks and stitching them with sensor numerical features into joint inputs), suitable for multiple inputs of machine learning models training.
- **Spatial and temporal alignment fusion:** This aligns the spatiotemporal coordinates of multimodal data (e.g., matches the PM2.5 concentration data from a weather station with the AOD (aerosol optical thickness) data inverted by satellites to the same geographic location and time point) and integrates complementary information through

weighted averages or model predictions (e.g., using machine learning models to establish the relationship between AOD and ground concentrations).
- **Semantic mapping:** This enhances cross-modality interpretability by defining uniform semantic labels for data in different modalities (e.g., associating "vegetation cover" in satellite imagery with "high biomass areas" in ecological observations).

Synergistic value of cleaning and pretreatment
Data cleaning is to "fix defects" and solve "errors and confusion" in the original data. Data preprocessing is an "optimized expression" that transforms clean data into "structured assets that adapt to analysis and visualization." Together, they form a key bridge between scientific data from "raw collection" to "usable information" — only after rigorous cleaning and reasonable preprocessing can scientific laws be accurately conveyed in subsequent visualizations (such as avoiding false trends caused by noise), support deep analysis (such as reliable training of machine learning models), and improve user experience (such as fluency and credibility during interactive exploration). In the following chapter, we further explore how this processed data can be used to achieve ultimate "scientific storytelling" through structured storage (e.g., database design and hierarchical organization) and visual toolchains (e.g., dynamic rendering and interaction design).

3.5. Data Structure for Digital Media Adaptation

After scientific data are acquired, cleaned, and preprocessed, they need to be organized and stored through a reasonable data structure to efficiently support visual rendering, interoperability, and multimodal fusion in the digital media environment. Digital media technologies (such as dynamic visualization, interactive exploration, and multimodal collaboration) put forward special requirements for data structures — not only to ensure the accuracy and timeliness of data but also to meet the needs of rapid retrieval, dynamic update, multi-dimensional correlation, and cross-modal mapping. The storage structure of traditional scientific data (such as simple tables or file collections) is often difficult to adapt to the complex interaction logic of digital media, so it is necessary to design a dedicated data structure for "digital media adaptation" to balance computing efficiency, storage cost, and user experience.

This section starts from the core requirements of digital media for data structures, analyzes and adapts to the typical data structure types of dynamic visualization, interactive perception, and multimodal fusion scenarios, and explains its design logic and typical applications in combination with the actual characteristics of scientific data.

3.5.1. *Core Requirements for Digital Media Adaptation: From Static Storage to Dynamic Interaction*

Scientific visualization in the digital media environment is no longer limited to "one-way transmission of static information" but emphasizes "user-driven dynamic exploration" and "multi-sensory collaborative information transmission." This shift presents four key requirements for data structures:

1. **Efficient retrieval and real-time response:** Users can request specific subset data (such as "temperature data in a certain area from 2020 to 2023" or "soil samples with a pH value greater than 7.5") through interactive operations (such as zooming, filtering, and dragging the timeline) in real time, and the data structure should support millisecond-level queries to avoid interaction lag caused by delays.
2. **Dynamic update and incremental processing:** For real-time monitoring data (such as second-level vibration signals from seismic networks and continuous temperature flow from industrial sensors), the data structure needs to support the rapid insertion of new data and the dynamic elimination of old data (such as sliding windows to retain the last 1 hour of data) while maintaining the stability of the overall structure.
3. **Multi-dimensional correlation and complex query:** Scientific data often contains multi-variables (such as temperature, humidity, and wind speed in meteorological data), multi-spatial–temporal scale (such as global grid + city site), and multi-modal information (such as image + sensor values), and the data structure needs to be able to efficiently express the correlation between variables (such as "coupling mode of temperature and humidity") and support cross-dimensional combination queries regions within a specific time period where temperature is >30°C and humidity is <60%.

4. **Cross-modal mapping and semantic association:** In multimodal fusion scenarios (such as correlating the pixel values of satellite images with the chemical concentration data of ground sensors), the data structure needs to explicitly record the semantic correspondence of different modal data (e.g., "the reflectivity of the red band of the image corresponds to the vegetation index NDVI") and support cross-modal joint retrieval (e.g., "Find NDVI> 0.8 and the corresponding area PM2.5 concentration <35 $\mu g/m^3$ point").

3.5.2. *Typical Data Structure Types: Adapt to Different Digital Media Scenarios*

In response to the above requirements, the digital media adaptation structure of scientific data can be divided into four categories: basic index structure, spatiotemporal dedicated structure, multimodal association structure, and dynamic streaming structure, each of which is optimized for specific interaction or visualization scenarios.

1. Basic index structure: "Navigation map" that accelerates single-dimensional retrieval

Applicable scenarios: These include scenarios where data need to be quickly filtered for a certain variable (such as time, spatial coordinates, and single parameter values) (such as "querying all observations within a certain date range" or "filtering soil samples with pH between 6.0 and 7.5").

Core structure and principles:

- **B-tree/B+ tree index:** This balances the multi-way search tree structure to support efficient range queries for ordered fields (such as timestamps and numeric parameters) (e.g., "Find data from September 1 to September 10, 2025"). The leaf nodes of the B+ tree are connected by a linked list, which is particularly suitable for range scanning (e.g., traversing all temperature records over a period) and is widely used in relational databases (e.g., MySQL and PostgreSQL) and scientific data management systems (e.g., NetCDF's variable index).
- **Hash index:** This implements a single point query with O (1) time complexity by mapping field values (such as sensor ID and sample

number) to the storage address through hash functions (e.g., "quickly obtain all data of sensors with ID 'SN001'"). The downside is that it doesn't support range querying (e.g., records with IDs between "SN001" and "SN010" cannot be found) for exact matching scenarios (e.g., unique identifier retrieval of experimental samples).

- **Bitmap index:** This creates a binary bitmap for each discrete value (e.g., soil type = sand/clay/loam) (1 means the sample belongs to the category, 0 means it does not belong) and quickly implements multi-conditional combination queries (e.g., "find samples with a type of sand and pH < 6.0") through bit operations (such as AND/OR). It is particularly suitable for low-base (small number of categories) categorical data and is commonly used for bioinformatics (e.g., genotypic classification) and remote sensing taxonomic data (e.g., land use type).

2. Space-time structure: "Spatial mesh" that handles multi-dimensional continuous fields

Applicable scenarios: Widely existing spatiotemporal continuous fields in scientific data (such as temperature grids of meteorological models, spatiotemporal propagation fields of seismic waves, and latitude and longitude distribution of ocean flow velocity) need to support efficient retrieval based on spatial location (such as "data from a certain latitude and longitude point") or time range (such as "data within the past 1 hour"), as well as aggregation at multiple spatiotemporal scales (such as down-sampling from a 1 km grid to a 10 km grid).

Core structure and principles:

- **R-tree/R* tree (spatial index):** This includes balanced tree structure, organizing the minimum bounding rectangle (MBR) of spatial objects (such as geographical grids, polygon areas) by hierarchy, and supporting efficient spatial range queries (e.g., "find all weather station data within Beijing") and nearest neighbor queries (e.g., "nearest seismic station within 10 km of a coordinate point"). R* trees are more suitable for dynamically updating spatial data (such as real-time traffic flow hotspots) than traditional R trees by optimizing insertion and splitting strategies.
- **Quadtree/octree (hierarchical grid):** Tree structure for recursively dividing space — quadtree for recursively dividing a two-dimensional plane into four quadrants (for planar maps) and octree for recursively dividing three-dimensional space into eight hexagrams (for volume

data, such as the 3D distribution of atmospheric temperature). Each node represents a subregion, and the data are stored in the corresponding leaf nodes according to location, supporting rapid spatial positioning (e.g., "find the average wind speed within a 10 km × 10 km grid") and multi-resolution analysis (e.g., step-by-step refinement from a coarse-grained grid to a target region).

- **Space-Time Cube:** The spatial dimension (latitude and longitude/plane coordinates) and the time dimension (timeline) are uniformly modeled as a three-dimensional cube, and each cell stores the data value of a specific spatiotemporal location (e.g., "The temperature at (116.4°E, 39.9°N) at 10:00 on September 6, 2025"). It supports spatial–temporal coupling analysis (e.g., "typhoon path moves with time") through slicing operations (such as viewing spatial distribution at a fixed time or viewing time changes at a fixed location) and is often used for dynamic continuous field data such as meteorology and traffic.

3. Multimodal association structure: Opening the "semantic bridge" of cross-source information

Applicable scenarios: Multimodal fusion data (such as satellite images + ground sensor data and medical images + genomic data) need to explicitly record the correspondence between different modal data (e.g., "a pixel of the image corresponds to a certain sampling point of the sensor") and support cross-modal joint retrieval and collaborative visualization.

Core structure and principles:

- **Graph:** Data entities (such as sensors, image blocks, and gene fragments) are represented as nodes, and the association relationships between entities (such as "sensor S1 is within the coverage range of image block I3" and "gene G1 interacts with protein P2") are expressed as edges (can be weighted, such as association strength or distance). The graph structure supports flexible multi-hop queries (e.g., "find all image regions related to an abnormal sensor data and their corresponding gene expression patterns") and can be stored and traversed efficiently through graph databases (such as Neo4j).
- **Metadata mapping table:** This explicitly records the correspondence of key fields of different modal data through a structured table (such as CSV or database table) (e.g., "the reflectance of the red band of the satellite image (Band 3) corresponds to the near-infrared band in the

NDVI calculation" and "the sampling time (UTC) of sensor S2 and the observation time of the weather station (UTC+8) require time zone conversion"). The table usually contains fields such as source modal ID, target modal ID, mapping rules (such as linear transformation formulas), and confidence level (such as associated error ranges), which can be manually used. Definitions or algorithms are automatically generated (e.g., matching images with sensor positions through spatial registration algorithms).

- **Unified identifier (UUID/GUID):** Each data entity (e.g., sample, observation, and image block) is assigned a globally unique identifier (e.g., UUIDv4) through which different modal data are associated (e.g., "Sample UUID=123456 corresponding sensor data UUID=789012 pH measurement"). This method decouples the specific field differences between modalities and only needs to establish a mapping relationship through identifiers, which is suitable for long-term storage and dynamic association of heterogeneous data sources (such as multi-source data integration in cross-agency cooperation projects).

4. Dynamic streaming structure: A "sliding window" for handling real-time data

Applicable scenarios: Real-time monitoring data (such as second-level vibration signals from seismic networks, continuous temperature streams from industrial sensors, and real-time public opinion data from social media) need to support the continuous influx of new data, automatic elimination of old data (such as retaining only the last 1 hour of data), and aggregate calculations in sliding windows (such as "average temperature in the past 5 minutes").

Core structure and principles:

- **Circular Buffer:** A fixed-size array structure that overwrites the oldest data when new data are written (pointer rotation), without the need for dynamic memory allocation, suitable for storing high-frequency data in a fixed time window (e.g., temperature sampling once per second, retaining the last 3,600 points or 1 hour of data). By maintaining the head and tail pointers and the current data volume, the insertion and elimination operations of O(1) time complexity can be realized.
- **Time-Series DB:** This is a database optimized for timestamped data (e.g., InfluxDB and TimescaleDB) that uses columnar storage (all variables of the same timestamp are stored in the same row) and

compression algorithms (e.g., delta encoding to reduce numerical differences at adjacent time points) to support efficient time-range queries (e.g., "Query all sensor data from 10:00:00 to 10:05:00 on 2025-09-06") and aggregation functions (e.g., mean and maximum values within the calculation window). Time-series databases usually have built-in expiration policies (such as automatic deletion of data older than 30 days) to adapt to the lifecycle management of real-time data.

- **Stream processing engine:** This is a combination of Apache Kafka (message queue) + Apache Flink (compute engine) to realize the "collection-buffer-processing-storage" pipeline of real-time data. New data first enter the message queue (e.g., Kafka's theme), and the stream processing engine calculates aggregated results (e.g., average temperature per minute) in real time according to rules (e.g., sliding windows and event time alignment) and writes the results to an adapted storage structure (e.g., time series database or graph database), supporting low-latency dynamic visualization (e.g., real-time updated temperature curves) (Ware, 2020; Sacha *et al.*, 2020).

3.5.3. *Adaptation Practice of Scientific Data: Take Typical Scenarios as Examples*

In order to understand the application of the above data structure more intuitively, the following is a combination of two typical scenarios of scientific visualization to illustrate the actual design logic of the digital media adaptation structure.

Scenario 1: Meteorological dynamic visualization (space-time continuous field + interactive exploration)
Requirements: Users can view the temporal and spatial changes of global or local temperature, wind speed, and other parameters in real time through interactive operations (zooming the map and dragging the timeline) and support multi-time comparison (e.g., "compare the temperature distribution of this year and the same period last year").

Adaptive structure design:

- **Spatial–temporal Cube + R-Tree Index:** This organizes global grid data (latitude and longitude×time) into spatial–temporal cubes, with each grid cell storing temperature values. Spatial scope queries are accelerated with R tree indexes (e.g., "find all grid points in the current view"), and time dimensions support timeline dragging (e.g., "quick locate to September 1, 2025") through B+ tree indexes.

- **Dynamic Flow Structure (Cyclic Buffer):** For real-time weather station data (such as ground temperature updated every minute), the cyclic buffer is used to store high-frequency data for the last 1 hour, allowing users to zoom in and view minute-level details in local areas. Historical data, such as climate data from the past 10 years, are stored in a space-time cube.
- **Multi-modal correlation (metadata mapping table):** If satellite cloud maps (image data) need to be superimposed, the correspondence between the spatiotemporal resolution of satellite images and the meteorological grid is recorded through the metadata mapping table (e.g., "the timestamp of the mesh corresponding to the satellite transit time, and the pixel coordinates of the image are mapped to the latitude and longitude grid") to realize the co-display of images and numerical data show.

Scenario 2: Biomedical Multimodal Analysis (Categorical Data + Cross-Modal Correlation)

Needs: Integrate patients' medical images (CT/MRI), clinical indicators (blood pressure, blood glucose), genomic data (SNP sites), and medical record text (symptom description) to support doctors through interactive screening (e.g., "find tumor volume>). 50 cm^3 and KRAS gene mutation") and correlated the lesion area and gene mutation site in the image.

Adaptive structure design:

- **Graph structure + unified identifier:** Each patient is the core node, and the associated sub-nodes include imaging data (UUID=IMG001), clinical indicators (UUID=BIO001), genetic data (UUID=GEN001), and medical record text (UUID=TEXT001). Edges represent associations (e.g., "IMG001 belongs to patients with BIO001"), and precise linking across modal data is achieved through unique identifiers (UUIDs).
- **Bitmap index (categorical variable) + B+tree index (numerical variable):** This uses bitmap index to accelerate multi-condition combination queries (e.g., "patients with adenocarcinoma and KRAS mutation") for categorical variables (e.g., tumor type = adenocarcinoma/squamous cell carcinoma, gene mutation status = wild-type/mutant

type). For numerical variables (e.g., tumor volume, blood glucose value), the B+ tree index supports range query (e.g., "tumor volume>50 cm^3").

- **Spatial–temporal index (R-tree, optional):** If the image data contain spatial information (such as the three-dimensional coordinates of the tumor in the CT image), the R-tree index can be used to accelerate the localization and visualization of the lesion area (e.g., "zoom in on the section of the lung lobe where the tumor is located").

Digital media adapts to the core value of data structure

The digital media adaptation structure of scientific data is essentially through the optimization of organizational methods, transforming "raw data" into "information carriers that can be efficiently interacted, deeply explored, and multi-modally integrated." From basic indexes to spatio-temporal-specific structures, from multimodal correlation to dynamic streaming processing, these structures together support the "dynamic," "interactive," and "immersive" of visualization in the digital media environment — they are not only tools for technical implementation but also a key bridge connecting scientific data and user cognition. In subsequent chapters, we further explore how to transform information in data structures into intuitive scientific stories through visual toolchains (e.g., dynamic rendering engines and interaction design frameworks) based on these structures.

Chapter 4

The Key Approach of Scientific Visualization: "Cognitive Translator" Based on Digital Media Technologies

4.1. Introduction: When Scientific Data Meets the Magic of Digital Media — A Transformation Engine from "Logic Coding" to "Visual Empathy"

Scientific visualization is essentially a "cross-language dialogue" — its "input" is the logical coding of scientific data (such as the exact magnitude of the number, the classification of variables, the order of time series, and the location correlation of space), "output" is visual information that the human brain can intuitively perceive (such as the depth and depth contrast of colors, the direction of flow of lines, the spatial structure of three-dimensional models, and the time-series evolution of animation). The human vision system is adept at capturing patterns, contrasts, and motion (such as spotting unusual patches of color and abrupt nodes in a fluid trajectory) but not directly understand abstract numerical meaning (e.g., "pressure value of 3.2 × 105 pa" or "fluctuation of variables at the 142nd time step") (Ware, 2020; Kitchin, 2022).

Digital media technology is the "translator" of this conversation — a comprehensive technological system that incorporates computer graphics, human–computer interaction design, and cognitive psychology. Be able to translate the logical coding of scientific data into visual elements that conform to the laws of human visual perception (e.g., map numerical sizes with coordinate axis locations and represent variable differences with color gradients) and through dynamic interaction (such as scaling, filtering, parameter adjustment) and multimodal presentation (such as the combination of audio prompts and three-dimensional immersion) to activate the user's pattern recognition instinct and ultimately from "data logic" to "visual resonance (Sacha *et al.*, 2020; Card *et al.*, 2022)."

If scientific data are a "crypto" to be interpreted, digital media technology is a key toolkit for decoding: It contains both basic "coding tools" (such as mapping quantitative data to point locations in a scatter chart and uses the high contrast of column charts to classify the values of variables) and also covers advanced "interactive tools" (such as displaying detailed parameters by mouse hover and dragging a slider to dynamically adjust the visualization dimension), and more involved advanced "immersion tools" (such as using virtual reality to "embed" users in 3D data fields, augmented reality to superimpose virtual graphs onto real scenes). The core goal of these tools is one — to make scientific data "alive," understandable, and exploreable so that researchers can discover hidden patterns, engineers can optimize designs, and the public can understand complex scientific concepts (Keim *et al.*, 2021).

This chapter focuses on the core approach of scientific visualization and systematically analyzes how to achieve this "cognitive translation" through digital media technology. We take a closer look along the technical lines of "Basic Coding — Dynamic Interaction — Immersive Rendering":

- How to select the correct visual channel (e.g., how to avoid misreading by color-blind users when distinguishing classification variables with color, how to ensure accurate perception when representing quantitative data with length).
- How to design composite encoding of multivariate data (e.g., when showing the coupling relationship of three variables, temperature, pressure, and flow rate, how to avoid visual confusion through layering of color, streamline, and transparency).
- How to activate deep exploration through interactive techniques (such as dynamic playback of time-series data, rotation and cutting of 3D models,and correlation analysis of multi-view interactions).

- How digital media features can be used to enhance cognitive experiences (such as simulating dynamic processes through animation and immersive observations at micro/macro scales with the help of virtual reality).

These methods are not only "how-to manuals" for technical implementation but also "strategy guides" for scientific visualization design — they provide guidance on how to use data characteristics (e.g., univariate/multivariate, static/dynamic, and micro/macro) and user goals (e.g., validating hypotheses, discovering patterns, and communicating knowledge) choose the best digital media technology solutions, ultimately allowing scientific data to "talk" through visual language as the core driver of scientific discovery and innovation practice.

4.2. Spatial Data Visualization

Spatial data are one of the most intuitive and exploratory types of scientific data, and their essence is the observation or simulation results associated with geographical coordinates (longitude, latitude, and elevation) or spatial location relationships (such as distance, direction, and adjacency), which are widely used in geography, geology, meteorology, ecology, urban planning, and other fields. From the perspective of scientific issues, spatial data often carries key information about "where" and "how" — e.g., the spatial distribution of earthquake epicenters reveals the activity law of plate boundaries, the location difference of urban PM2.5 concentration reflects the correlation between industrial layout and population density, and the gradient change of ocean surface temperature correlates with the dynamic characteristics of ocean current systems. Digital media technologies (such as dynamic rendering, interoperation, and 3D modeling) provide powerful tools for the visualization of spatial data, upgrading it from a traditional "static map" to a scientific cognitive carrier that can be explored, analyzed, and immersed (MacEachren, 2021; Andrienko *et al.*, 2020).

This section focuses on the core methods of spatial data visualization, systematically analyzes its data characteristics, typical visualization forms, digital media enhancement technology, and scientific application scenarios, and reveals how to transform spatial data into intuitive scientific insights through digital media technology (Sheppard, 2021; Dykes *et al.*, 2020).

4.2.1. *Core Characteristics and Classification of Spatial Data*

The key difference between spatial data and non-spatial data is its strong position-related properties, which include both absolute position (such as the specific latitude and longitude coordinates of a weather station) and relative position (such as the distance between two sampling points or the upstream and downstream relationships between monitoring points in the basin). According to the spatial dimension and data structure, spatial data can be divided into the following three categories:

1. Point data

Definition: It represents a spatial location through discrete coordinate points (single or multi-point collections), each associated with one or more measurements (e.g., temperature, concentration, and event count).

Typical scenarios: Mineral deposit sampling points (coordinates + ore grade) in geological exploration, case occurrence location (GPS coordinates + onset time) in infectious disease monitoring, and single-point weather station data (latitude and longitude + temperature/pressure) in meteorological observation.

Key attributes: Location accuracy (e.g., GPS error range) and statistical significance of attribute values (e.g., single-point values represent local features and regional laws need to be inferred from density analysis).

2. Line data

Definition: It is a linear feature connected by a series of consecutive coordinate points to represent a spatial object with a direction or path feature, and each line or segment can be associated with attributes (such as flow, speed, and resistance).

Typical scenarios: River network (river section coordinates + flow/flow velocity), traffic network (road centerline + vehicle speed/congestion index), geological fault line (fault plane coordinates + displacement), and animal migration trajectory (flight/walking path composed of GPS continuous points).

Key attributes: Topological relationship (such as connecting nodes and intersections of lines), directionality (such as the upstream and downstream directions of rivers, positive and negative electrode flow directions of current), and segmentation characteristics of line elements (different line segments may be associated with different attribute values).

3. Polygon data

Definition: It is a polygonal area enclosed by a closed line to represent a spatial extent with continuous coverage features, with each face-related attribute (such as average, total, and category).

Typical scenarios: Administrative division (provincial/municipal boundary + population density), land use type (coverage of farmland/forest/urban land), ocean basin (sea basin boundary + average depth), and ecological reserve (boundary coordinates + biodiversity index).

Key attributes: Regional closure (to ensure that there are no gaps in geometric closure), internal homogeneity assumption (intraplane attribute values are usually regarded as relatively uniform, but there may be local differences in practice), and adjacency relationships (such as the interaction between adjacent planes, such as the cross-regional diffusion of urban heat island effects).

Special type: Raster data

Although raster data are strictly classified as a "regular grid spatial continuous field," they are often included in the category of spatial data because they are widely used to express spatial distribution (such as the pixel matrix of satellite remote sensing images and the grid point elevation of digital elevation model DEM). Raster data consist of regular 2D/3D grid cells, each of which (pixels/voxels) stores a numerical value (e.g., reflection, rate, temperature, and altitude) and indirectly represents the position through the matrix index of the grid, which is suitable for expressing continuously changing fields (such as the gradient of air temperature with space).

4.2.2. *Typical Visualization Forms: From Static Mapping to Dynamic Interaction*

Digital media technology provides a rich choice of forms for the visualization of spatial data, which can be divided into the following five core methods according to the dimension of data (2D/3D), dynamics (static/dynamic), and interaction requirements (passive observation/active exploration):

1. Point Density/Discrete Mapping

Method logic: The discrete point data are thermally mapped by symbol size, color, or density to its attribute values or spatial aggregation features, and the "where it is concentrated and where it is sparse" is visually displayed.

Digital media enhancements:

- **Symbol mapping:** This uses points of different shapes/colors to distinguish attribute categories (e.g., earthquake points are divided into colors according to magnitude, case points are divided into symbols by infection type) and combines sizes to code numerical strength (e.g., the higher the grade of the mineral deposit, the larger the point).
- **Density heat map:** Kernel Density Estimation (KDE) calculates the aggregation intensity around points (e.g., points per unit area) and represents density gradients with gradient colors (e.g., blue → yellow → red) to reveal implicit distribution patterns (e.g., urban crime hotspots).
- **Dynamic interactions:** They support zooming (to view local aggregation details), filtering (to filter points of a specific category by attribute value), and hovering (to display details of individual points).

Scientific applications: They include mineralization intensity distribution at geological sampling points, range markers of wildlife habitats, and spatiotemporal spread trajectories of epidemic cases.

2. Linear feature mapping

Method logic: This encodes the value of a line by geometric properties (e.g., length and curvature) and visual variables (e.g., color, width, and transparency) or dynamic processes (e.g., flow changes and path directions).

Digital media enhancements:

- **Static line diagram:** This uses color gradients to represent changes in attributes along the line (e.g., the water temperature of a river rises from upstream to downstream, using blue → green → red gradient) and maps the traffic size with width (e.g., the larger the traffic flow on the main traffic road and the thicker the line).
- **Dynamic flow effect:** This simulates dynamic processes (such as ocean current direction, air flow trajectory) through particle animation (e.g., water flow particles in a river moving along a river section) or streamline (e.g., ocean current direction and air flow trajectory), combined with velocity mapping (particle movement speed indicates flow velocity) to enhance intuitiveness.
- **Interoperative:** This supports clicking online segments to view attribute details (such as real-time congestion index of a certain road), dragging to adjust the streamline density (controlling the clarity of animation), and segment filtering (focusing on line features in a specific section).

Scientific applications: These include hydrological simulation results of river networks, load distribution of power transmission lines, and seasonal changes in animal migration paths.

3. Polygonal mapping

Method logic: It is encoding a face's attribute values (such as averages, categories) or spatial relationships (such as contrasts of adjacent areas) by its fill color, texture, or boundary style.

Digital media enhancements:

- **Choropleth map:** This divides the area into several levels according to the attribute value (e.g., the population density is divided into low/medium/high), and uses different colors (such as light green → dark green → red) to represent the level difference, and visually compares the relative size between areas (pay attention to the confusion of "area-attribute," and small areas with high values may be covered by large areas with low values).
- **Proportional symbol map:** This superimposes symbols proportional to attribute values (such as circles/squares, and size represents total GDP) in the face, combines the color of the surface to distinguish

categories (such as different industry types), and conveys the total volume and classification information at the same time.
- **3D terrain/body visualization:** This extends the area data to 3D (such as digital elevation model, DEM, generates undulating terrain and superimposes land use surfaces) and improves spatial perception through lighting effects (shadows enhance stereoscopic sense), transparency adjustment (displays underground structures), or dynamic profiles (cuts surfaces display internal attribute distribution).
- **Interactive exploration:** This supports click highlighting of surface elements (focusing on specific areas), hovering to display statistical summaries (such as the average temperature and extreme values of an administrative region), and multi-level zoom (level by level refinement from country → province → city).

Scientific application: This includes comparison of population/economic indicators of administrative divisions, regional distribution of ecosystem service functions, and risk assessment of geological disaster susceptibility zoning.

4. Raster Field Mapping

Method logic: For raster data (such as satellite remote sensing images, digital elevation models, and meteorological temperature fields), the numerical distribution and change trend of continuous space are presented by color mapping, contour/contour, and dynamic rendering.

Digital Media Enhancements:
- **Color mapping:** This converts the values of raster cells to color by predefined color scale → s (e.g., temperature with blue (cold), white (room temperature), → red (hot) gradients), combined with pseudo-color (artificial color scale enhancement contrast for non-visible light bands, e.g., infrared radiation) to highlight key features;
- **Equal Value Line/Equal Value Surface:** This extracts lines (equal value lines, such as topographic contour lines) or surfaces (equal value surfaces, such as regions above a specific temperature threshold) where the values are equal. By adjusting the thickness of the lines/transparency of the surfaces, key gradients can be highlighted (e.g., areas with dense isobar lines in meteorology indicate strong wind belts).
- **Dynamic temporal rendering:** This shows dynamic evolution (such as typhoon generation and dissipation process) through animations

(such as daily satellite cloud maps and hourly temperature fields) or timeline sliders (manually select moments) for multi-temporal raster data (e.g., daily satellite cloud maps and hourly temperature fields).

- **Multi-band fusion:** This combines multiple raster bands (such as red, green, and blue bands to synthesize true color images, and near-infrared bands superimpose vegetation index NDVI) into multi-dimensional information carriers, combined with interactive band selection (user-defined display band combinations) to support multi-parameter analysis.

Scientific applications: These include surface temperature monitoring of satellite remote sensing, spatial distribution of marine chlorophyll concentration, and three-dimensional rendering of seismic wave velocity models.

5. 3D Spatial Visualization

Method logic: This expands spatial data from 2D plane to 3D (three-dimensional) space (or adds time dimension to form 4D) and encodes more information through additional dimensions such as height, depth, and volume to support more realistic scene restoration and immersive exploration.

Digital Media Enhancements:

- **3D terrain modeling:** This generates real terrain surfaces based on digital elevation models (DEM), superimposes surface coverage (e.g., vegetation and buildings) and point/line elements (e.g., sensor locations and roads), and enhances three-dimensionality through dynamic lighting (sun azimuth changes) and shadows.
- **Volume Rendering:** This uses ray casting or maximum density projection (MIP) algorithms for 3D raster data (such as air pollutant concentration fields and medical CT scans) to directly display the attribute distribution (e.g., concentration gradient of haze in building complexes) within the body.
- **Dynamic interaction:** This supports perspective rotation (observing the spatial structure from any direction), zooming (focusing on local details or macroscopic global), flight roaming (dynamically browsing the scene along a preset path), and section cutting (cutting the three-dimensional body through a virtual plane to display internal cross-sectional properties).

- **Multimodal fusion:** This combines image data (such as satellite image mapping to the terrain surface), sensor data (such as marking the real-time value of monitoring points in 3D terrain), and audio prompts (such as playing an alarm sound when approaching a high-risk area) to enhance immersion.

Scientific applications: These include simulation of pollution diffusion in urban three-dimensional space, three-dimensional structural analysis of geological bodies (such as the spatial relationship between faults and folds), and stereo dynamic process display of ocean/atmospheric circulation.

4.2.3. *The Core Enhancement Role of Digital Media Technology*

Compared with traditional paper maps or static charts, digital media technology brings the following key breakthroughs to spatial data visualization:

- **Dynamics:** This supports "process visualization" through animations (e.g., time-series changes) and real-time updates (e.g., dynamic rendering of sensor streams) to reveal implicit dynamic patterns (e.g., real-time tracking of typhoon paths and diurnal changes in urban heat island effects).
- **Interactivity:** Users can actively explore data through zooming, filtering, hovering, clicking, and other operations (such as focusing on high-value points in a specific area and comparing polygon attributes in different time periods), from "passive reception" to "active discovery."
- **Multi-dimensional correlation:** This combines multiple visual variables such as color, size, height, and time to display multiple attributes at the same time (such as temperature + humidity + population density in a certain area) or correlate spatial and non-spatial data (such as superimposing the analysis of case distribution and medical resource distribution).
- **Immersion:** 3D visualization and virtual reality (VR)/augmented reality (AR) technologies (such as superimposing geological faults onto field scenes through AR glasses) are possible. Users are "immersive"

in perceiving spatial relationships, improving cognitive intuitiveness and memory depth.

The scientific value and practical logic of spatial data visualization:

The essence of spatial data visualization is to transform abstract "data values" into intuitive "spatial patterns" through the visual coding of location associations, from the resource enrichment zone found by point aggregation, to the energy transmission path revealed by line flow, to the environmental gradient changes reflected by surface differences, and finally to restore the complex correlation of the real world through three-dimensional structure. The integration of digital media technology not only enhances the expressiveness and interactivity of visualization but also promotes the leap from "data display" to "scientific discovery": Scientists can quickly verify hypotheses through dynamic interaction (such as "whether the confluence pattern of tributaries in a watershed affects flood peaks"), and the public can understand complex scientific problems (such as "how global warming causes sea level rise and coastline erosion") through immersive experience. In the following chapters, we further explore the visualization methods of non-spatial data (such as time series and network) and the visualization strategies of multimodal data fusion.

4.3. Temporal Data Visualization

Temporal data are one of the most common types of scientific data, and their core characteristic is that the data values are strongly associated with the time dimension — each data point is accompanied by an explicit timestamp (e.g., second, minute, hour, day, or year) that records how a variable has changed over time. From single-variable time series (such as daily temperature at a place) to multi-variable time sequences (such as simultaneous monitoring of temperature, humidity, and wind speed), from regular sampling (such as sensor data recorded every 10 minutes) to irregular sampling (such as event-driven observation records), time-series data exist widely in the fields of astrology, economics, biology, engineering, and so on, carrying the key scientific information of "when" and "how." Digital media technologies (such as dynamic rendering, interactive manipulation, and multivariable synchronous presentation) provide powerful tools for time-series data visualization, upgrading it from a traditional "static line chart" to a dynamic analysis vector that is "explored, predictable, and correlated."

This section focuses on the core methods of time-series data visualization, systematically parses its data characteristics, typical visualization forms, digital media enhancement techniques, and scientific application scenarios, and reveals how digital media techniques can transform time-series data into intuitive scientific insights (Iliinsky, *et al.*, 2011; Spence, 2020).

4.3.1. *Core Features and Classifications of Time-Series Data*

The key dimension of time-series data is "timing," but its specific form is significantly different from the way relevant to the problem in science. According to the time dimension, the number of variables, and the sampling characteristics, time-series data can be divided into the following categories:

1. Classification By Temporal Granularity

- **High-frequency time-series data:** These include short sampling intervals (second to minute) and are often used to capture dynamic processes that change rapidly (e.g., voltage fluctuations per second on the grid, millisecond vibration signals from seismometers, and high-frequency financial transaction data).
- **Low- and low-frequency time-series data:** These include sampling at a longer interval (hours to grade level) to describe relatively slow trends (e.g., daily temperature, monthly GDP growth rate, and annual species populations).

2. Classification By Number of Variables

- **Single-variable time-series data:** These contain only a time series of a target variable (e.g., daily PM2.5 concentration in a city and body temperature monitoring record of a patient), focusing on analysis of the variable's own patterns (e.g., cyclical and trend).
- **Multivariate time-series data:** These include time series containing multiple related variables (such as temperature, humidity, wind speed, air pressure) that need to be analyzed for coupling between the variables (e.g., "How coordinated changes in temperature and humidity affect the index of sensation").

3. Classification By Sampling Rules

- **Regular time-series data:** These include fixed sample intervals (e.g., meteorological data recorded every hour) and uniform distribution of time axes facilitate direct application of traditional time series analysis methods.
- **Unregular time-series data:** Sampling intervals are not fixed (e.g., seismic magnitude recorded only at the time of an event and physiological indicators when a patient develops symptoms), and time alignment issues need to be solved by inserting values or special processing methods.

4.3.2. *Typical Forms of Visualization: From Basic Mapping to Multi-dimensional Association*

The time-series data visualization method supported by digital media technology can be divided into the following five core forms according to the analytical objectives (trend identification, anomaly detection, and pattern association) and the interaction requirements (dynamic exploration and multivariate comparison) (Tufte, 2022; Cleveland, 2021):

1. Basic Timeline Plot

Methodology: This uses time as a horizontal axis (X axis) and variable values as a vertical axis (Y axis) and connects the data values of each time point through continuous line segments (cosine chart) or discrete points (scatter chart) to visualize the trend of variables over time.

Digital media enhancement:

- **Dynamic rendering:** This supports dynamic propulsion of the timeline (such as displaying data points from left to right step by step and simulating a real-time monitoring process), or controlling the time window by a slider (manually selecting the range of time periods to observe).
- **Multi-Variable Superposition:** This include drawing multiple lines in the same graph (different color/line differentiators) to compare the simultaneous variation of multiple related variables (e.g., temperature and amount of precipitation).
- **Interaction:** Hover to display the values of a specific time point (tools prompts), click to zoom in on a local time period (focus on key

changes), and drag and zoom to adjust the time resolution (switch from annual view to day view).

Scientific applications: These include the long-term trend of global annual average temperature, the seasonal fluctuation of a river flow during the year, and the recovery process of patients' postoperative vital signs (heart rate/blood pressure).

2. Periodic and Trend Decomposition Plot

Method logic: Based on the theory of time-series analysis (e.g., the classical additive model Y _ t = T _ t + S _ t + R _ t, where T _ t is the trend term, S _ t is the season term, and R _ t is the residual term), the time-series data are decomposed into different components and their contributions are displayed by hierarchical visualization.

Digital media enhancement:

- **Multi-subplot synchronization:** This displays raw time-series graphs, trend lines (smoothing curves), seasonal fluctuations (cyclic repeating patterns), and residues (random noise) and associates the corresponding time points by color/transparency.
- **Dynamic labeling:** This automatically identifies and labels significant cycles (e.g., annual cycles in meteorological data, day-night cycles) or turning points in trends (e.g., inflection points for economic growth and critical times for ecosystem degradation).
- **Interaction validation:** Users can observe changes in decomposition results in real time by adjusting decomposition parameters (such as smooth window size) to verify the validity of cycle assumptions.

Scientific applications: These identify inter-year variations and long-term warming trends in climate data, analyze seasonal spikes and long-run growth demands of power loads, and detect trends in equipment aging (often increasing disability) in industrial sensor data.

3. Multivariate Temporal Correlation

Methodology: This shows the time-series changes of multiple related variables at the same time and reveals the coupling or lag effect between the variables through visual coding (e.g., color, shape, and dynamic correlation) (e.g., "Humidity begins to decline 2 hours after temperature increases").

Digital media enhancement:

- **Matrixed layout:** The time series of multiple variables is arranged in matrices (rows and columns), the diagonal line shows the details of the single variable, and the non-diagonal line shows the strong or weak correlation between the variables (e.g., Pearson correlation coefficient).
- **Synchronous brightening:** When the user selects a time point or variable, the corresponding periods of other related variables are automatically brightened (e.g., selecting a high temperature moment and synchronizing brightening of the periods of low humidity and high evaporation at the same time).
- **Dynamic lag analysis:** This adjusts the amount of time shift by a slider (e.g., "moves the axis of variable B forward/back N hours") and observes the changes in correlation between variables in real time (finds the optimal lag time for the maximum correlation).

Scientific applications: These study the mechanisms of interaction of meteorological factors (temperature, humidity, and air pressure), analyze unusual correlation patterns of multisensory monitoring data (such as multiphysics field parameters for earthquake precursors), and explore coordinated fluctuations of economic indicators (GDP, unemployment, and consumption index).

4. Event Driven Timeline (Event-Annotated Timeline)

Methodology: This attaches time tags to the underlying time-series diagram for key events (e.g., policy implementation, natural disaster, and experimental operations) and highlights the effect of events on the time-series data through special symbols (e.g., vertical lines and icons) or color changes.

Digital media enhancement:

- **Event library association:** A database of events (including event name, time, type, and impact description) is established, and users can click on event tags to see detailed information (e.g., "After the implementation of epidemic prevention and control policies in March 2020, traffic in a city plunged by 30%").
- **Comparative views:** These provide pre-event–post-event segmental comparisons (e.g., by splitting the timeline by a dash, showing the mean/variance changes before and after an event separately) or animate the dynamic response of data after an event is triggered.

- **Interaction filtering:** This supports filtering by event type (e.g., showing only natural disaster-related events) or local zoom analysis that focuses on a specific event period.

Scientific applications: Evaluating the short-term and long-term impacts of policy interventions on environmental indicators such as carbon emissions and air quality; analysis of ecosystem response processes before and after major natural disasters (e.g., earthquakes or floods), and study of the immediate effects of experimental procedures (e.g., drug injections or equipment start-up) on biological/physical parameters.

5. 3D Temporal Surface
Methodology: Using time as a third dimension (Z-axis), combined with traditional X (time)–Y (variable values) two-dimensional graphs, a curved or body rendering displays complex time-series patterns with multiple variables or time periods (e.g., "Three-dimensional evolution of temperatures in different regions over time").

Digital media enhancement:
- **Dynamic rotations and profiles:** These support viewing angle rotation of three-dimensional curves (observing time-series patterns from any direction) and horizontal/vertical profiles cutting (showing spatial distribution at a moment or time-series curves at a location).
- **Color and height encoding:** The curved height represents the size of the variable value and the color map represents another variable (such as the temperature curve uses the height to represent the temperature value and the color to represent the humidity value), enabling multi-variable synchronous encoding.
- **Timeline animation:** This visualizes the accumulation or evolution of time-series data by dynamically playing the generated curve (e.g., overlaying an hourly layer of data frame by frame) (e.g., how snow depth increases with wintertime).

Scientific applications: These include multi-region, multi-temporal temperature precipitation combined evolution of global climate models, seasonal changes in the three-dimensional structure of ocean circulation, and stress–time–spatial responses of engineering structures (e.g., bridges) under long-term loads.

4.3.3. *Core Enhancements of Digital Media Technologies*

Compared to traditional static time-series graphs, such as paper charts or single-frame line charts, digital media technologies have brought the following key breakthroughs for time-series data visualization:

- **Dynamic:** This is through animation (e.g., a timeline progressively, data points appearing one by one) or real-time updates (e.g., the dynamic rendering of a sensor stream), uncovering dynamic processes in time-series data (such as "hourly increases in the intensity of typhoons" compensates for the failure of static graphs to show "processes of change").
- **Interaction:** Users can zoom (focus on a critical period), filter (select a specific variable or event), and hover (view accurate values). Operations such as drag (adjusting the time window) actively explore data, moving from "passive reception" to "active discovery" of hidden patterns (such as "prior signal before an exceptional value appears").
- **Multivariate correlation:** This shows the time-series changes of multiple related variables (e.g., temperature + humidity + wind speed) simultaneously, through color, shape, dynamic correlation, and other encodings, revealing coupling relationships between variables (e.g., whether a negative correlation between humidity and temperature is stronger in a particular season) to support more complex scientific hypothesis validation.
- **Immersive experience and analytics:** Combining 3D visualizations (e.g., 3D sequential surfaces) with VR (e.g., "walking into" the view of space to observe multiple dimensions), increase the intuitive perception of complex time-series relationships (e.g., "spatial–temporal co-evolution of multi-area temperature gradients").

The Scientific Value and Practical Logic of Time-Series Data Visualization

The essence of time-series data visualization is to transform abstract "numerical changes" into intuitive "dynamic processes" through visual coding of time dimensions — identifying cyclical patterns (such as the four-season temperature cycle) from single-variable trends. Identify coupling mechanisms from multivariate associations (e.g., interactions between meteorological factors) and evaluate the effects of interventions from the impact of events (e.g., policy regulation of the environment).

The integration of digital media technology not only improves the presentation and interactivity of visuals, further advancing the leap from "data recording" to "process understanding" to "predictive derivation": Scientists can validate hypotheses (such as whether the collapse of an ecosystem is related to continuous drought events) through dynamic interactions. The problem in science can be visualized to understand how global warming is affecting daily life through year-to-year temperature increases. In subsequent chapters, we further explore spatial–temporal co-visualization strategies (e.g., "the three-dimensional evolution of a region's temperature over time and space") and visualization strategies for unstructured temporal–temporal data (e.g., time tags in text logs).

4.4. Multi-dimensional Data Visualization

Multi-dimensional data are one of the most complex and information-dense types of scientific data, and their core feature is that a single data entity (such as a sample or an observation point) contains multiple interrelated variables (dimensions) at the same time, and these variables may belong to different physical dimensions, statistical types (continuous/discrete/categorized), or semantic categories (such as gene expression in biology, protein structure parameters, and synergy of environmental factors). In meteorology, an observation point may record more than ten meteorological elements such as temperature, humidity, air pressure, wind speed, and precipitation at the same time. In the field of biomedicine, a patient's medical record data may contain dozens or even hundreds of variables, such as age, gender, gene sequence, biochemical indicators, and imaging characteristics. In industrial production, a device's sensor network may collect continuous parameters such as temperature, vibration, current, and pressure in real time, along with discrete or categorized information such as equipment model, operating duration, and maintenance records (Maaten & Hinton, 2021; van der Maaten, 2022).

The challenge with multi-dimensional data is that human perception is limited (often only focusing on 3–4 separate dimensions at once), and scientific problems often require unearthing hidden patterns, associations, or anomalies from dozens of variables. Digital media technologies (such as dimensionality reduction projection, parallel coordinates, interactive filtering, and multi-view collaboration) provide powerful tools for spatial data visualization, allowing scientists to "see" otherwise invisible high-dimensional patterns by compressing high-dimensional information into 2D/3D perceptible space, or by displaying complex relationships between variables through multi-view linkage. This section focuses on the core methods of multi-dimensional data visualization, systematically analyzes its data characteristics, typical visualization forms, digital media enhancement technologies, and scientific application scenarios, and reveals how to transform multi-dimensional data into explorable scientific insights through digital media technology.

4.4.1. *Core Features and Challenges of Multi-dimensional Data*

The core dimensions of cube data are "number of variables" and "variable types," and their complexity is reflected in the following aspects:

- **High dimensionality:** The number of variables usually exceeds the limit of direct human perception (e.g., meteorological data contain 10–20 meteorological elements, and gene expression data contain transcript abundance values of tens of thousands of genes), resulting in "dimensionality curse" — as the dimension increases, the distance between data points tends to be uniform, the reliability of traditional statistical methods (such as clustering and classification) decreases, and the correlation patterns between variables in visualization are difficult to visualize.
- **Heterogeneity:** Variables may belong to different physical dimensions (e.g., temperature (°C), pressure (Pa), or concentration (mol/L)) and statistical types (e.g., continuous variables (e.g., wind speed), discrete variables (e.g., equipment status: normal/faulty), or categorical variables (e.g., soil type: sandy/clay/loam)), which need to be standardized or coded for uniform comparison.
- **Complex correlations:** Nonlinear relationships between variables (e.g., exponential correlation between temperature and biological activity), redundancy (e.g., multiple sensors measuring different components of the same physical quantity), and implicit patterns (e.g., certain combinations of variables show significant correlations only under certain conditions) can be revealed by advanced analytical methods (e.g., principal component analysis, clustering, and association rule mining).

4.4.2. *Typical Visualization Forms: From Dimensionality Reduction Projection to Multi-View Collaboration*

The multi-dimensional data visualization method supported by digital media technology can be divided into the following five core forms according to the analysis objectives (overall pattern recognition, variable association exploration, and anomaly detection) and interaction requirements (dynamic screening, multi-dimensional linkage):

1. Dimensionality Reduction Projection

Method logic: This includes mapping high-dimensional data (N-dimensional) to low-dimensional space (usually 2D or 3D) through mathematical algorithms, preserving the relative distance or structural relationship between data points, to display the clustering, outlier, or distribution patterns of high-dimensional data in a perceptible 2D/3D space.

Common algorithms and visualization logic:

- **Principal Component Analysis (PCA):** This is a linear dimensionality reduction method that converts the original variables into a set of unrelated principal components (sorted by variance contribution rate) through orthogonal transformation and selects the first 2–3 principal components (covering most of the data variations) to map to 2D/3D space. For example, 10 elements (temperature, humidity, air pressure, etc.) in meteorological data may be compressed into 2 principal components (PC) by PCA1 and PC2), where PC1 represents "comprehensive temperature-pressure change" and PC2 represents "humidity-wind speed correlation," and the clustering distribution of data points in the PC1-PC2 two-dimensional graph can reveal the patterns of different weather types (such as sunny, cloudy, and rainy data clusters).
- **t-SNE (t-Distributed Stochastic Neighbor Embedding):** This is a nonlinear dimensionality reduction method that focuses on preserving local structures in high-dimensional data (e.g., similar data points remain close to each other in low-dimensional space) and is suitable for fine clustering of data (e.g., local aggregation of different cell types in gene expression data). However, t-SNE has a weak retention of the global structure (the relationship between distant data points may be distorted) and is usually used in combination with PCA (first PCA is

used to reduce dimensionality to about 50 dimensions, and then t-SNE is further compressed to 2 dimensions).

- **UMAP (Uniform Manifold Approximation and Projection):** This is another nonlinear dimensionality reduction method, which has better adaptability to the global structure while retaining the local structure, and has higher computational efficiency than t-SNE, making it suitable for large-scale multi-dimensional data (such as high-dimensional feature sets of tens of thousands of samples).

Digital Media Enhancements:

- **Dynamic Shading:** This colors low-dimensional projection points based on the value of a raw high-dimensional variable (e.g., "Precipitation intensity" in meteorological data) or cluster labels (e.g., categories grouped by K-means) to reveal the relationship between the variable and the dimensionality reduction structure (e.g., "data points with high precipitation intensity are concentrated on the positive semi-axis of PC1").
- **Interactive exploration:** This supports hovering to display original high-dimensional data values (tooltip), clicking to highlight similar data points (focus on specific clusters), and dragging to rotate low-dimensional projections (adjusting the viewing angle).
- **3D projection:** This maps the dimensionality reduction results to 3D space (such as the first three principal components of PCA) and observes the data distribution more comprehensively through interactive operations such as rotation and scaling (such as discovering spirals or layered structures in 3D space).
- **Scientific applications:** These include pattern classification of meteorological elements (such as data clusters that distinguish frontal rain, convective rain, and terrain rain), cell type identification of gene expression data (t-SNE clustering of different cell subpopulations), and customer behavior analysis (dimensionality reduction grouping of e-commerce user purchase characteristics).

2. Parallel Coordinates Plot

Method logic: Each dimension (variable) is represented as a parallel vertical axis (coordinate axis), and each data point is represented as a polyline connecting the corresponding values of each axis (Keim *et al.*, 2021; Sacha *et al.*, 2020).

Visualization principle:

- Each vertical axis represents a variable (e.g., temperature, humidity, and wind speed), and the scale on the axis represents the value range of the variable (e.g., temperature 0–40°C).
- Each data point (e.g., meteorological observation record at a certain moment) is connected by a polyline to its numerical position on each axis (e.g., 25°C on the temperature axis, 60% on the humidity axis, and 5m/s on the wind speed axis).
- Clustering or separation patterns of multiple polylines reflect coupling relationships between variables (e.g., "temperature and humidity usually change in reverse" lines show a crossover pattern) or clusters of data (similar polylines clustering to form "bunches").

Digital Media Enhancements:

- **Dynamic filtering:** This achieves multi-dimensional linkage screening by clicking on a range of a certain axis (e.g., "temperature >30°C") or dragging to highlight the polylines that meet the conditions and automatically highlights the corresponding values on other axes (e.g., after filtering out the high temperature data, the corresponding low humidity polyline on the humidity axis will also be marked).
- **Cluster shading:** This visualizes the characteristic patterns of each cluster (e.g., a clustered polyline with a specific slope on the temperature–wind speed axis) based on pre-calculated clustering results (e.g., K-means grouping) to color the polylines (e.g., different colors represent different clusters).
- **Dimension reordering:** This allows users to manually adjust the order of axes (e.g., arranging closely related variable axes next to each other) to enhance the intuitiveness of pattern recognition (e.g., placing the "temperature" and "dew point temperature" axes close to each other, making it easier to observe the linear relationship between the two).

Scientific applications: These include variable correlation analysis of multi-sensor monitoring data (such as synchronous changes in vibration, temperature, and current of industrial equipment), comprehensive evaluation of multiple indicators in social science research (such as correlation patterns of economic development, education level, and environmental quality), and clustering of sample characteristics in bioinformatics (such as the combination of clinical indicators of patients with different diseases).

3. Radar Chart/Star Plot

Methodology: This maps the values of multiple variables (typically 3–8) onto an axis that radiates outward from a central point (like a spider web structure). Each data point is represented as a closed polygon (radar chart) or a polygonal star (star chart). By comparing the shape, area, and vertex positions of the polygons, the comprehensive performance of different samples on multiple-dimensional variables can be assessed.

Visualization principle:

- Each axis represents a variable (e.g., height, weight, arm length, leg length, and lung capacity) of the organism, and the scale of the axis is uniformly standardized (e.g., normalized to 0–1) to ensure comparability.
- The value of each data point (such as an athlete or a species) on each axis is converted into the position of the corresponding vertices, connecting all vertices to form a closed graph.
- After the radar map of multiple data points is superimposed, the advantages and disadvantages of each sample in multi-dimensional variables or pattern differences are visually compared through the overlap, separation, and shape differences of graphs (e.g., athlete A is prominent in the speed and strength axes, and athlete B is dominant in the endurance and flexibility axes).

Digital Media Enhancements:

- **Dynamic comparison:** This supports displaying radar maps of multiple data points at the same time (such as comparison of patient indicators in different treatment groups) and avoids visual confusion through transparency adjustment or color differentiation.
- **Interactive highlighting:** This clicks on a data point to highlight its graph while fading other graphs (focusing on a specific sample) or hover to display specific values for each variable.
- **Variable filtering:** This dynamically adjusts the dimensions of the radar map by checking/unchecking axes (variables) (e.g., retain only key indicators to simplify complex graphics).

Scientific application: This includes comprehensive evaluation of biological characteristics (such as the comparison of multiple traits of different varieties of plants), multi-index performance analysis (such as the comprehensive score of finance, market, and innovation in enterprise operations), and pattern recognition of medical indicators (such as the multi-dimensional distribution of physiological parameters of patients with different diseases).

4. Scatter Plot Matrix (SPLOM)

Method logic: This combines all the variables in the cube data in pairs to generate a matrix of N×N (N is the number of variables); each cell in the matrix is a two-dimensional scatter plot (showing the correlation of the two variables), and the diagonal cells usually display the distribution histogram or density graph of the single variable and show the full correlation pattern between the variables through the matrix as a whole.

Visualization principle:

- The rows and columns of the matrix represent variables in the multi-dimensional data (e.g., temperature, humidity, air pressure, and wind speed).
- The scatter plot of cells (i,j) shows the pairwise relationship between variables i and variable j (e.g., scatter distribution of temperature and humidity) and judges the coupling strength and direction of the two through the aggregation pattern of points (e.g., linear correlation, no correlation, and clustering).
- Diagonal cells (e.g., $i = j$) show the distribution of variable i itself (e.g., histogram of temperature values) to help users understand the basic statistical characteristics of a single variable (e.g., skewness and kurtosis).

Digital Media Enhancements:

- **Dynamic highlighting:** This selects a specific data point (such as high temperature value) in one cell and automatically highlights the corresponding associated points in other cells (such as highlighting the humidity and wind speed values corresponding to the high temperature point in the humidity-wind speed cell at the same time) to achieve cross-variable correlation tracking.
- **Statistical annotation:** This superimposes correlation coefficients (such as Pearson r-value) or regression lines in scatter plot cells to

quantitatively display the strength of linear relationships between variables.

- **Interactive filtering:** This dynamically updates the association pattern in the matrix by filtering data points in all cells simultaneously through global filter conditions (e.g., “temperature >30°C”).

Scientific applications: These include multivariate correlation analysis (such as statistical correlation between meteorological elements), variable selection of feature engineering (such as screening input features strongly related to target variables in machine learning), and multi-factor coupling research of ecosystems (such as the comprehensive effects of temperature, precipitation, and soil nutrients on plant diversity).

5. Multi-View Coordination

Method logic: This combines multiple visualization forms (such as dimensionality reduction projection map, parallel coordinate map, radar map, and scatter plot) in one interface and synchronizes the view linkage and interaction to realize the correlation mode of synchronizing the exploration of multi-dimensional data from different angles.

Typical combination and interaction logic:

- **Dimensionality reduction projection + parallel coordinates:** The dimensionality reduction map shows the global clustering structure of the data (such as the data cluster in the 2D PCA graph), and the parallel coordinate map simultaneously displays the contribution of each dimension to the clustering (e.g., clicking on a cluster in the dimensionality reduction graph, and the polyline of the corresponding data point in the parallel coordinate map is highlighted) highlighting key variables.
- **Radar map + scatter plot matrix:** The radar map focuses on the comprehensive comparison of a few key variables (such as 3–4 core traits of different species), and the scatter plot matrix shows the full correlation of all variables (assists in verifying whether the patterns in the radar map are universal).
- **3D scatter plot + parallel coordinates:** 3D scatter plot visually displays the spatial distribution of three core variables (such as 3D clustering of temperature, humidity, and air pressure), and parallel coordinate map supplements the association of other secondary variables (such as the influence of wind speed and precipitation).

Digital Media Enhancements:

- **Cross-view highlighting:** This select data points in one view (such as an abnormal sample), and the corresponding points in other views will be automatically highlighted (e.g., outliers in the dimensionality reduction map will also be marked with corresponding polylines in the parallel coordinate map).
- **Unified filtering:** This filters data in all views simultaneously through the global filter panel (e.g., "temperature >30°C and humidity <60%") to maintain consistency across views.
- **Dynamic layout:** This supports user-defined view arrangement and size (such as enlarging the key parallel coordinate map and shrinking the secondary radar map) to optimize exploration efficiency.

Scientific applications: These include multi-dimensional comprehensive analysis of complex scientific problems (such as the synergistic impact of temperature, precipitation, ecological indicators, and human activities in climate change research), multimodal data fusion (such as the joint diagnosis of medical imaging characteristics and clinical indicators), and multi-parameter fault diagnosis of industrial systems (such as comprehensive analysis of equipment vibration, temperature, current, and pressure signals).

4.4.3. *The Core Enhancement Role of Digital Media Technology*

Compared with traditional static cubic tables or single-dimensional charts, digital media technology brings the following key breakthroughs to multi-dimensional data visualization:

- **Dimensionality reduction and compression:** High-dimensional information is mapped to 2D/3D perceptible space through PCA, t-SNE, and other algorithms, breaking through the limitations of human perception dimensions and making hidden clustering or outlier patterns visible (e.g., "originally scattered data points are aggregated into specific categories after dimensionality reduction").
- **Interactive exploration:** Users can actively focus on specific dimensions or subsets of data (e.g., "only observe samples with a temperature >30°C and wind speed >5 m/s") by filtering, highlighting, dragging, etc., and move from "passive reception" to "active discovery" of

correlations between variables (e.g., "negative correlation between humidity and temperature is only significant in summer").

- **Multi-view collaboration:** The combination and linkage of multiple visualization form supports all-round analysis from global (such as the overall structure of dimensionality reduction projection) to local (such as the distribution details of univariates), from correlation (such as coupling mode between variables) to anomalies (such as feature combinations of outliers).
- **Dynamic updates and real-time:** For multi-dimensional data monitored in real time, such as industrial sensor networks or environmental monitoring stations, digital media technology supports dynamic rendering and real-time filtering (e.g., "update data every minute and automatically adjust clustering results") to assist in real-time decision-making (e.g., "trigger an alarm when an abnormal data cluster is detected").

The scientific value and practical logic of multi-dimensional data visualization:

The essence of multi-dimensional data visualization is to transform the "invisible mode" in high-dimensional space into an "intuitive structure" in low-dimensional space through dimensionality reduction projection, multi-view linkage, and interactive exploration — from the clustering mode of meteorological elements to the feature combination of biological samples, from the anomaly detection of industrial parameters to the correlation analysis of socio-economic indicators, digital media technology enables scientists to break through the limitations of human perception and dig out key clues that drive scientific discovery from complex multi-dimensional data. In the following chapters, we further explore the visualization strategies of multi-dimensional temporal data (e.g., "multi-dimensional patterns of variables with time") and spatial and multi-dimensional combined data (e.g., "multi-dimensional environmental indicators in different regions"), promoting the expansion of scientific visualization to higher-dimensional and more complex scenarios.

4.5. Network Data Visualization

Network data are special and complex types of scientific data, and their core feature is a non-Euclidean structure composed of nodes and edges — nodes represent entities (such as individuals, organizations, molecules, and neurons), and edges represent relationships between entities (such as social interactions, chemical reactions, neural connections, and traffic routes). From the social network of users on the Internet to the protein interaction network in organisms, and from the connection of roads and stations in the transportation system to the co-authorship of scholars in scientific research cooperation networks, network data exist widely in social sciences, biology, computer science, traffic engineering and other fields, carrying key scientific information of "who is connected to whom" and "how strong and what pattern." Digital media technologies (such as force-oriented layout, dynamic interaction, and multi-level aggregation) provide powerful tools for spatial data visualization, allowing scientists to "see" the community structure, key nodes, and dynamic evolution processes in the network by transforming abstract network structures into intuitive graphic images (Jacomy *et al.*, 2021; Fortunato, 2020).

This section focuses on the core methods of network data visualization, systematically analyzes its data characteristics, typical visualization forms, digital media enhancement technologies, and scientific application scenarios, and reveals how to transform network data into explorable scientific insights through digital media technology.

4.5.1. *Core Characteristics and Classification of Network Data*

The core dimension of network data is the relationship between "nodes" and "edges," and its complexity is reflected in the number of nodes, the connection mode of edges, and the global and local structure of the network. According to the attributes of nodes and edges, and the structural characteristics of the network, network data can be divided into the following categories:

1. Classification by Type of Node and Edge

- **Undirected network:** The edges are not directional and only represent the binary association between nodes (e.g., user A and user B are friends in a social network, and there is no need to distinguish between "A follows B" or "B follows A"). Typical examples include scientific research cooperation network (paper co-authored by Scholar A and

Scholar B) and transmission lines in power networks (nodes are substations, and edges are line connections).

- **Directed network:** The edges are directional, indicating a one-way relationship between nodes (e.g., in a social network, User A pays attention to User B, but User B may not pay attention to User A; neuron A in the neural network transmits signals to neuron B). Typical examples include web link networks (hyperlinks from source pages to destination pages) and predatory relationships in the food chain (predators → predators).
- **Weighted network:** The edge comes with a weight value that indicates the strength or capacity of the relationship (e.g., the frequency of interaction between user A and user B in a social network (number of messages per day) and the traffic capacity of roads in the transportation network (traffic per hour)). Weights can be continuous values (e.g., interaction intensity from 0 to 100) or categorical levels (e.g., strong/medium/weak associations).
- **Attributed network:** In addition to the connection relationship, the node or edge also has additional attribute information (such as the age, gender, and occupation of the user in the social network; molecular function and subcellular localization of proteins in protein networks). Attribute information provides richer semantic context for network analysis (e.g., "young scientists are more inclined to interdisciplinary collaboration").

2. Classification by Network Size and Structural Complexity

- **Small-scale simple networks:** The number of nodes is small (e.g., dozens to hundreds), and the edge connection pattern is relatively simple (e.g., sparse joining or regular mesh), which is usually used for fine analysis of local structures (e.g., the cooperation mode of small scientific research teams).
- **Large-scale and complex networks:** These include the number of nodes (e.g., thousands to hundreds of millions) and dense edge connections (e.g., global social networks or Internet web links) and may contain multiple relationships (e.g., multiple types of interaction between users: likes, comments, and retweets), and the analysis of their global structure (e.g., community distribution and central nodes) and local characteristics (e.g., neighborhood relationships of key nodes) requires efficient visualization and computational methods.

4.5.2. *Typical Visualization Forms: From Basic Layout to Dynamic Interaction*

The network data visualization method supported by digital media technology can be divided according to the analysis objectives (overall structure identification, key node detection, and community discovery) and interaction needs (dynamic exploration and multi-level analysis). The following are the five core forms:

1. Force-Directed Layout

Method logic: Based on the "force model" (such as spring force and repulsive force) in physical simulation, the nodes are regarded as charged particles (mutual repulsion) and the edges are regarded as springs (attraction of connecting nodes), and the position of the nodes is iteratively calculated to achieve the mechanical equilibrium state as a whole, so as to visually display the local connection mode and global structure of the network.

Visualization principle:

- The repulsive force between nodes (such as Coulomb force) makes the nodes as dispersed as possible to avoid overlapping.
- The elasticity of the edges (such as Hooke's law) keeps the directly connected nodes as close as possible to maintain the tightness of the connection.
- The sparsity or compactness of the layout (e.g., highlighting local clusters or global connections) can be controlled by adjusting the parameters of the force (e.g., repulsive strength and spring rate).

Digital Media Enhancements:

- **Dynamic interaction:** This drags nodes to adjust the layout (manually optimize the position of key nodes), clicks nodes to highlight their neighbors (focus on local connections), and zooms and pans to observe the global or local structure.
- **Edge visualization optimization:** This adjusts the line thickness (the higher the weight, the thicker the line) or transparency (the lower the weight, the lighter the line) or reduces the edge intersection with curves (such as arcs) (improve readability).
- **Node coding:** This includes degree (the number of connected edges and reflecting centrality), color to indicate the node's category

(e.g., user type in a social network), or attribute value (e.g., molecular function classification of proteins).

Scientific applications: These include user interaction structure of social networks (e.g., identifying opinion leaders and community groups), protein interactions in biomolecular networks (e.g., core proteins of signaling pathways), and hub stations in transportation networks (e.g., key positions of airports and subway transfer stations).

2. Community Detection and Clustering

Method logic: Algorithms (such as Louvain algorithm, Girvan–Newman algorithm, and modularity optimization) are used to identify a subset of nodes (i.e., "communities" or "clusters") with tight internal connections and sparse external connections in the network, and visually distinguish different communities through color, shape, or boundary lines to reveal the global organizational structure of the network.

Visualization principle:

- Nodes within communities usually have more edge connections (high density) and fewer node connections between communities (low density).
- Modularity is an indicator that measures the quality of community division (the higher the value, the more significant the community structure), and the algorithm optimizes community division by maximizing modularity.
- When visualizing, each community fills the node background with different colors or draws bounding boxes, allowing users to quickly identify groups with different functions or semantics (such as "family community" and "colleague community" in social networks).

Digital Media Enhancements:

- **Dynamic community adjustment:** This supports users to manually merge or split communities (such as merging adjacent small communities into one large community) and observes changes in modules in real time.
- **Cross-community connection highlighting:** This displays key connection edges between communities (such as the edge with the highest weight or the edge connecting different community core nodes) to reveal the interaction pattern between communities.

- **Hierarchical display:** For nested community structures (such as large communities containing small communities), multi-scale organizational structures are displayed through multi-level layouts (such as concentric circles or tree nesting).

Scientific applications: These include disciplinary communities in research cooperation networks (e.g., interdisciplinary research groups in physics and computer science), functional groups of species in ecological networks (e.g., pollinator communities and plant communities), and topic communities on the Internet (e.g., user clusters around specific topics in social media).

3. Key Nodes and Centrality Analysis

Method logic: By calculating the centrality indicators of nodes (e.g., degree centrality, intermediate centrality, proximity centrality, and eigenvector centrality), identify the nodes in the network that play the most critical role in information propagation, structural stability, or function (e.g., "super-communicator" in social networks and "bottleneck nodes" in transportation networks), and highlight them through visual coding (e.g., size, color, and label).

Common central indicators and visualization logic:

- **Degree centrality:** This is the number of directly connected edges (number of neighbors) of a node, the higher the degree of the node, the more "central" the node (e.g., the one with the most friends in the social network and household).
- **Betweenness centrality:** This is the frequency with which nodes appear in all the shortest paths (acting as a "bridge" to control the transmission of information), and nodes with high intermediary numbers are usually key connection points between communities (e.g., transfer hubs in transportation networks).
- **Closeness centrality:** This is the reciprocal of the average shortest path length from the node to all other nodes (reflecting the efficiency of information arrival), and the node with high closeness centrality is closer to the center of the network.
- **Eigenvector centrality:** This is considering not only the number of direct neighbors of a node but also the importance of neighbors (e.g., "people connected to important people are also more important"), often

used to identify "influential nodes" in the network (e.g., highly cited scholars in academic networks).

Digital Media Enhancements:

- **Multi-indicator overlay coding:** This uses node size to represent degree centrality and color to indicate intermediary centrality (e.g., red = high intermediary and blue = low intermediary) or directly displays the values of key indicators through labels.
- **Dynamic filtering:** Users can filter nodes by centrality threshold (e.g., only the top 10% of the key nodes of the mediation centrality) to focus on the core structure of the network.
- **Path tracing:** This supports clicking on a key node to highlight the shortest path to other nodes (such as the most likely route for information propagation).

Scientific applications: These include identification of "super spreaders" in disease transmission networks (public health intervention targets), protection of key substations in power networks (prevention of cascade failures), and analysis of core scholar cooperation patterns in academic cooperation networks.

4. Dynamic Network Evolution

Method logic: For dynamic networks that change over time (such as the increase or decrease of user relationships in social networks with time and the gene interactions of biological networks with developmental stages), the evolution process of network structure (addition, deletion, and weight change of nodes/edges) is shown through timeline animation or time slice sequences, and the dynamic characteristics of the network (such as the merger and division of communities or the rise and fall of key nodes) are revealed (Brandes & Raab, 2021; Vehlow *et al.*, 2022).

Visualization principle:

- **Timeline drive:** This arranges multiple time snapshots of the network (e.g., monthly/yearly network status) in chronological order and shows continuous changes in the network through animation (frame-by-frame switching) or slider control (manually select time points).
- **Incremental update:** This only renders new or deleted nodes/edges (e.g., when a new user joins a social network, only add the

corresponding nodes and edges, instead of redrawing the entire network) to improve dynamic rendering efficiency.

- **State comparison:** This superimposes networks at multiple points in time (such as the network showing the previous moment with a translucent layer) and visually compares structural differences (such as "a community expanded from 3 nodes to 10 nodes in 3 months").

Digital Media Enhancements:

- **Dynamic interaction:** This supports pausing/playing animations, adjusting time steps (such as switching from "yearly" to "monthly" to observe more granular changes) and clicking on specific points in time to view detailed network status.
- **Event correlation:** This correlates the time markers of external events (such as policy releases, environmental changes) with network evolution (e.g., "after the implementation of a regulation, the weights of some edges in the network increase significantly") to assist in causal analysis.
- **Multivariate synchronization:** This displays the changes of node/edge attributes over time (e.g., the weight of the edge indicates the transaction amount, and the line thickness in the animation is dynamically adjusted with the increase or decrease of the amount).

Scientific applications: These include user relationship dynamics of social media (such as the rapid formation and dissipation of user communities caused by hot events), changes in gene regulatory networks during biological development (such as gene interaction networks at different stages of embryonic development), and evolution of Internet web links (such as the rise of new themes and the decline of traditional websites).

5. Multi-Level Network

Method logic: For networks containing hierarchical structures (such as hierarchical relationships in organizations, such as "senior managers → middle departments → grassroots employees," and "protein complexes → individual proteins → amino acids) in biological networks, the association patterns of nodes and edges at different levels are shown through nested layouts (such as concentric circles or tree maps) or hierarchical rendering, and the synergy between macro and microstructure is revealed.

Visualization principle:

- **Hierarchical division:** This divides nodes into multiple levels (e.g., Layer 1 is macro entities, Layer 2 is sub-entities, and Layer 3 is more granular components), and edges can be connected across levels (e.g., macro entities and children entities are directly related, or only within the same level).
- **Nested layout:** This uses concentric circles to represent the hierarchy (the outer layer is macro, the inner layer is microscopic), the position of the nodes reflects their belonging in the hierarchy, and the edges are connected across the hierarchy and the same level by color or line type.
- **Hierarchical rendering:** This displays the high-level structure (such as the division of departments in the organizational structure) first and then gradually expands the sub-levels (such as clicking on the department node to display its subordinate teams and members), supporting a step-by-step exploration mode.

Digital Media Enhancements:

- **Hierarchy switching:** Users can choose to focus on a specific level (e.g., only show connections between macro entities) or expand to a multi-level joint view (observe macro–micro interactions).
- **Cross-level interaction:** When clicking a high-level node, automatically highlight its associated sub-level nodes (e.g., after clicking "Research Institution," highlight its subordinate "laboratory" and "researcher").
- **Aggregation and expansion:** Aggregation of micro-level nodes (such as merging multiple similar grassroots nodes into a "cluster") simplifies the display of complex networks and allows users to expand details on demand.

Scientific applications: These include management structure and collaborative relationship in enterprise organizations (such as the impact of high-level decision-making on grassroots execution), multi-level regulatory network in organisms (such as the transmission path of hormone signals from cell membrane receptors to nuclear transcription factors), and macro road network and micro station connection in urban transportation (such as the relationship between subway lines and station passenger flow).

4.5.3. *The Core Enhancement Role of Digital Media Technology*

Compared with traditional static network diagrams such as hand-drawn schematics or simple node-edge lists, digital media technology brings the following key breakthroughs to network data visualization:

- **Layout optimization:** Through physical simulation methods such as force-oriented algorithms, abstract network structures are transformed into intuitive 2D/3D graphics (the spatial position of nodes and edges reflects the connection relationship), breaking through the limitations of human cognition of complex topologies (e.g., "dense networks that are otherwise difficult to understand clearly display community grouping through layout").
- **Interactive exploration:** Users can actively focus on specific substructures (such as local communities or key nodes) by dragging nodes, clicking and highlighting, zooming and panning, and moving from "passive reception" to "active discovery" of hidden patterns in the network (e.g., "a small community is actually a bridge between multiple large communities").
- **Multi-dimensional correlation:** This includes combining the attributes of nodes/edges (such as categories, weights, and centrality indicators) and dynamic changes (such as time evolution), through the synchronous encoding of multiple visual variables such as color, size, and animation, to support all-round analysis from "structure" to "function" to "dynamic" (e.g., "key nodes are not only the connection center, but also assume the role of information hub in a specific period of time").
- **Dynamic and multi-layered:** For dynamic and multi-level networks, digital media technology supports timeline animation and nested layouts, allowing scientists to observe the evolution of networks over time (e.g., how the merger and division of communities affect overall stability) or macro–microstructure synergies (e.g., how high-level decisions are transmitted to the executive layer through multi-layered structures).

The scientific value and practical logic of network data visualization:

The essence of network data visualization is to transform the abstract "relational network" into an intuitive "structural image" through the graphical coding of nodes and edges — from the community interaction mode in social networks to the signal transduction pathways in biological networks, from the hub connection structure in the transportation network to the interdisciplinary mode in the scientific research cooperation network, digital media technology enables scientists to break through the cognitive barriers of complex topologies and dig out the key clues that drive scientific discoveries from the global organization and local details of the network (such as "how the failure of key nodes causes cascading failures" and "how dynamic changes in the community reflect social trends"). In the following chapters, we further explore the visualization strategies of multimodal networks (such as nodes and edges containing multiple types of data at the same time) and hyperscale networks (such as Internet graphs with hundreds of millions of nodes), to promote the expansion of network science visualization to more complex and dynamic scenarios.

4.6. Generative Science and Creative Visualization

In traditional scientific visualization, the mapping relationship between data and graphs is often clear and fixed — e.g., temperature data re visually represented by color gradients (blue → red), and time series are presented by continuous trajectories of line graphs. However, as scientific research extends from "data description" to "knowledge discovery," "science dissemination," and even "integration of science and art," the demand for visualization by scientists and the public is no longer limited to "accurately transmitting data" but further pursues "inspiring," "revealing underlying patterns" and "creating immersive experiences." Generative science and creative visualization are the response to this need: The former automatically generates scientifically meaningful visualizations (such as motion graphics simulating natural phenomena and data-driven artistic creation) through algorithms, while the latter emphasizes the transformation of scientific data into "scientific stories" with both cognitive value and aesthetic experience through artistic expression, interactive narrative, and cross-modal integration. Digital media technologies (such as generative adversarial network GANs, procedural generation, and interaction design) provide key technical support for this process, upgrading scientific visualization from a "tool" to a "bridge connecting science and humanities (Goodfellow *et al.*, 2020; Hertzmann, 2021)."

This section focuses on the core methods of generative science and creative visualization, systematically analyzes its scientific foundation, typical technical paths, digital media implementation strategies, and cross-domain application scenarios, and reveals how to transform scientific data into more enlightening visualization results through generative logic and creative expression.

4.6.1. *Core Connotations of Generative Science and Creative Visualization*

The essence of generative science and creative visualization is to break through the single mapping of "data-fixed graphics" in traditional visualization through algorithm-driven automatic generation and artistic expression design and realize the multi-dimensional expansion from "data presentation" to "knowledge exploration," "science communication," and "aesthetic experience." Its core characteristics can be summarized as follows:

1. **Data-driven generation logic:** Based on scientific data (e.g., observations, simulation results, or multimodal parameters), algorithms (e.g., procedural rules or machine learning models) automatically generate visual graphics or dynamic effects instead of relying on manual design. The generated content not only retains the scientific authenticity of the data (such as the accuracy of values and the objectivity of relationships) but also expands the expression dimension of the data through the creativity of algorithms (such as simulating the dynamic process of natural phenomena and generating multi-perspective data metaphors).
2. **Integration of art and science:** This integrates art design principles (such as color aesthetics, compositional balance, and narrative rhythm) into scientific visualization, enhancing the cognitive appeal and emotional resonance of visualizations through visual metaphors (such as using flowing rivers to metaphorically compare data flows and mapping high-temperature regions with flame colors), interactive narratives (such as scientific discovery stories that follow the timeline), or cross-modal expressions (such as combining sound and tactile feedback to enhance immersion).

3. **Dynamic and interactive generation process:** Generative visualization is usually dynamic (such as simulated graphics that evolve over time) or interactive (such as user actions triggering different generation results), allowing scientists or the public to actively participate in the generation of visual content through exploration behaviors (such as adjusting parameters and selecting perspectives) to discover hidden patterns in the data or generate personalized scientific expressions.

4.6.2. *Typical Technical Paths: From Algorithm Generation to Artistic Expression*

The realization of generative science and creative visualization relies on the synergy of multiple technical paths, which can be divided into three categories: data-driven programmatic generation, machine learning-assisted generation, and interactive generative design, each of which targets different scientific goals and creative needs.

1. Procedural Generation

Technical logic: By writing program rules (such as mathematical functions and algorithmic logic), scientific data (such as parameter values and spatial–temporal coordinates) are mapped into graphic elements (such as points, lines, and surfaces, textures), and visual structures with specific scientific significance are automatically generated. At the heart of procedural generation is "rule definition" — scientists control the form, dynamics, and interaction behavior of generated graphics by setting parameters and logical relationships.

Typical Methods and Applications:

- **Physics simulation-based generation:** This utilizes physics engines (e.g., fluid dynamics simulations, particle systems) to generate natural laws. The data input are physical parameters (such as temperature, pressure, and wind speed), and the output is a realistic dynamic visualization. For example, in meteorology, the distribution of clouds under different atmospheric circulation modes is simulated through procedural generation, which helps researchers intuitively understand the dynamic mechanism of the climate system.

- **Parametric graph generation:** This converts scientific data (e.g., multivariate parameters) into geometric figures (e.g., complex curves and 3D surfaces) through mathematical functions (e.g., fractal geometry and Fourier transform). For example, the molecular structure data can generate a three-dimensional atomic model through parameterization rules, in which the atomic spacing is determined by the bond length data and the molecular configuration is controlled by the bond angle parameters. In geology, fractal algorithms are used to generate simulated stratigraphic fold morphology to assist researchers in analyzing the complexity of actual geological structures.
- **Procedural rules for data mapping:** This maps the numerical range of scientific data (e.g., temperature 0–100°C) to graphical attributes (e.g., color gradient (blue → red), height (0–100 pixels), and texture density (sparse → dense)) and automatically generates color maps, heightmaps, or texture maps through program logic. For example, in ecology, the species diversity index is mapped to the shade of color on the map through programmatic rules to quickly identify biological hotspots.

2. Machine Learning-Assisted Generation

Technical logic: This uses machine learning models (such as generative adversarial network GANs, variational autoencoder VAE, and diffusion models) to learn the intrinsic distribution or patterns of scientific data and automatically generate visualization content with data characteristics (such as synthetic images simulating real observation data and generating multi-perspective scientific concept maps). The advantage of machine learning models lies in their ability to capture complex nonlinear relationships in data, such as implicit coupling between high-dimensional parameters, to generate diverse results that go beyond traditional rule definitions.

Typical Methods and Applications:

- **Generative Adversarial Networks (GANs):** These generate realistic scientific data visualization results (e.g., synthetic images of satellite cloud images and simulation of cell morphology under a microscope) through adversarial training of generators and discriminators. For example, in astronomy, GANs can learn the distribution characteristics of real galaxy images and generate simulated galaxies with different morphologies (spirals and ellipses) and brightness distributions, helping researchers verify the generalization ability of galaxy classification models. In medical imaging, GANs can generate synthetic images of pathological sections (such as cell distributions of different cancer types) to train diagnostic models or popularize disease characteristics to the public.

- **Variational autoencoder (VAE) and diffusion model:** This generates visual variants with data semantics (e.g., different styles of data charts and multi-perspective scientific concept diagrams) by learning the latent space of the data. For example, VAE can compress high-dimensional meteorological data (such as multivariable combinations of temperature, humidity, and wind speed) into low-dimensional latent space, and users can adjust the parameters of the latent space (such as sliders to control the "temperature-humidity" weight) to generate dynamic visualization effects in different climate modes (such as rainstorm cloud maps corresponding to "high temperature and high humidity" and clear sky maps corresponding to "low temperature and low humidity"). Diffusion models can gradually generate animations of complex scientific processes (such as molecular collision trajectories of chemical reactions and morphological evolution of biological development) and show the details of dynamic processes by controlling the number of steps (such as time steps).
- **Style transfer and cross-modal generation:** This combines scientific data with artistic styles (such as oil paintings, watercolors, and science fiction concept maps) to generate visual works with both scientific information and artistic beauty (such as converting satellite remote sensing images into van Gogh-style night maps of the Earth, highlighting the distribution of urban lights) and generates protein structure data into cyberpunk-style 3D models for interactive displays at science exhibitions.

3. Interactive Generative Design

Technical logic: Through real-time interaction between users and the visualization system (such as parameter adjustment, perspective switching, and gesture operation), personalized visual content is dynamically generated so that the expression of scientific data changes with user needs or exploration behavior. At the heart of interactive generation is "user-driven" — scientists or the public no longer passively receive fixed visualizations but actively participate in the generation process and discover data through active actions, hidden patterns, or generating scientific expressions that align with specific goals (Segel & Heer, 2021; Burch & Weiskopf, 2022).

Typical Methods and Applications:

- **Parametric interaction control:** This provides an interactive interface (such as sliders, drop-down menus, and color pickers), allowing users to adjust key parameters of the generation rules (such as color mapping range, time step of dynamic simulation, and number of fractal iterations) and observe the dynamic changes of the generated results in real time. For example, in ecological simulation, users can generate population distribution maps under different ecological conditions by adjusting parameters such as "species reproduction rate" and "environmental carrying capacity" and intuitively understand the regulation mechanism of ecological balance. In meteorological visualization, users can generate customized weather warning maps by modifying parameters such as "wind speed threshold" and "precipitation probability."
- **Multimodal interaction generation:** This combines multiple interaction methods (such as voice commands, gesture recognition, and eye tracking) with multimodal feedback (such as sound, touch, and vision) to create an immersive generative experience. For example, in a science exhibition, visitors can adjust the three-dimensional perspective of the molecular structure by swiping gestures and at the same time hear the sound feedback of the vibration frequency of the corresponding chemical bonds (high frequencies indicate strong bonds, low frequencies indicate weak bonds). In medical education, medical students can select different anatomical structures such as "heart valves" and "vascular networks" through voice commands, systematically generate corresponding 3D visualization models, and superimpose dynamic blood flow simulation effects.
- **Story-driven generation:** This is the process of combining scientific data with interactive narratives (e.g., timeline stories and question-answer processes) to gradually generate a visualization of scientific findings through the user's exploring behavior (e.g., clicking "Next" and "Deeper Inquiry"). For example, in a popular science app, users can review "the evolution of human understanding of the universe" through an interactive timeline — from the ancient core of the Earth to the modern theory of the Big Bang, each click generates a corresponding scientific visualization (such as the Copernican map of the solar

system, the Hubble pictures of galaxies), while showing the notes and quotes of key scientists, enhancing the narrative and emotional resonance of science communication.

4.6.3. *The Core Supporting Role of Digital Media Technology*

The realization of generative science and creative visualization relies on several key capabilities of digital media technologies that provide the underlying support for algorithm generation, artistic expression, and interaction design:

- **High-performance computing and real-time rendering:** Generating complex dynamic visualizations (such as fluid simulations, particle systems, and 3D molecular models) requires powerful computing power (such as GPU parallel computing) and real-time rendering technologies (such as OpenGL and WebGL) to ensure the smoothness of the generation process and the realism of graphics.
- **Artificial intelligence and machine learning frameworks:** Machine learning models such as generative adversarial networks (GANs), variational autoencoders (VAEs), and diffusion models rely on deep learning frameworks (such as TensorFlow and PyTorch) to learn data distribution and generate scientifically significant visualizations by training large-scale scientific datasets (such as satellite images, microscope images, and meteorological parameters).
- **Interaction design and user experience (UX):** By designing intuitive interactive interfaces (such as parameter sliders and gesture operation prompts), multimodal feedback (such as voice/haptic prompts), and narrative flows (such as question guidance and step-by-step exploration), users can participate in the generation process more conveniently and interestingly, making scientific visualization more accessible to non-professional groups.
- **Cross-media and cross-modal integration:** This combines visual (graphics and animation), auditory (sound and music), touch (vibration and force feedback), olfactory (simulating environmental odors), and other multimodal expressions to create immersive scientific experiences (such as scientific scene roaming in virtual reality, VR, and data superposition display in augmented reality, AR) and expands the expression dimension and emotional influence of scientific visualization.

4.6.4. *Cross-Field Application Scenarios: From Scientific Research Exploration to Science Communication*

Generative science and creative visualization have shown unique value in many fields, and their application scenarios can be divided into three categories: scientific research support, science education and public communication, and science and art integration.

1. Scientific Research Support: Explore hidden patterns in your data
In scientific research, generative visualization serves as a tool for scientists to explore complex data, validate hypotheses, or simulate experiments. For example, in astronomy, GANs are used to generate synthetic galaxy images to assist researchers in testing the robustness of galaxy classification algorithms. In biology, programmatic molds are generated, which simulate cell growth morphology under different environmental conditions to provide theoretical prediction for experimental design. In climate science, parametric interactive control is used to generate global temperature distribution maps in multi-climate modes, which helps researchers understand the long-term impact of greenhouse gas emissions on climate change.

2. Science Education and Public Communication: Enhance the Interest of Scientific Cognition
In science education and public communication, generative visualization transforms abstract scientific concepts into intuitive and interesting visualizations through artistic expression and interactive design, stimulating learning interest and public participation. For example, in the popular science application for primary and secondary school students, the "three-state change of water" is displayed through interactive generation (users adjust the temperature parameters and generate corresponding ice, water, and steam dynamic graphics in real time). In the science museum, visitors can generate a 3D model of the "human blood circulation system" through gesture operations and hear the sound feedback of blood flow. On social media, scientists use GANs to generate concept maps of "the impact of climate change on future cities" (such as coastal buildings after sea level rise) to artistically convey scientific evidence of the environmental crisis to the public.

3. Integration of Science and Art: Creating Innovative Expressions Across Disciplines

Generative science and creative visualization promote the deep integration of science and art (such as digital art, design, and new media art) to create works that combine scientific rigor and artistic beauty. For example, the artist collaborates with climate scientists to transform global carbon emissions data into dynamic "Earth Breathing" visualizations through programmatic generation — changes in carbon dioxide concentrations are mapped to the flow of light and shadow on the Earth's surface, visualizing the impact of human activities on the Earth's system. Biologists and digital designers use VAE to generate concept maps of "future bio-morphism" (based on genetic data and evolutionary trends of real species) to explore the potential possibilities of biodiversity. Physicists collaborate with new media artists to artistically convey abstract physical concepts by showcasing the visual metaphor of "quantum entanglement" (such as the synchronized movement of two interconnected particles in virtual space) through interactive generation.

Future prospects for generative science and creative visualization

The core value of generative science and creative visualization lies in transforming scientific data from "objective record" to "subjective exploration," "knowledge transmission," and "aesthetic experience" through the creative expansion of algorithms and the expressiveness of art. In the future, with the maturity of generative artificial intelligence (such as scientific narrative generation driven by large language models) and mixed reality (MR) technology, generative science and creative visualization will further break through the limitations of two-dimensional screens, develop in the direction of immersive, personalized, and interdisciplinary, and become a "visual bridge" connecting scientists, the public, and the natural world. In the following chapters, we further explore the technical implementation details of generative visualization (such as algorithm selection and parameter tuning) and cross-disciplinary collaboration processes (such as the collaboration model between scientists and artists), promoting this emerging field from theoretical exploration to practice.

Chapter 5

Interaction and Immersive Science Visualization

5.1. Introduction: When Visual Interfaces Become "Science Laboratories" — How Interaction and Immersion Reshape Cognitive Dimensions

In the traditional scientific visualization mode, the user often acts as a "passive observer" (pillar graphs) or 3-D models with fixed viewing angles, where they can infer data patterns only indirectly from limited visual information (such as color, depth, and graphic size). When you need to delve deep into details (e.g., zooming in on a specific region and comparing different time steps) or adjust the analytical dimension (e.g., switching combinations of variables and modifying parameter conditions), you must rely on limited interaction features preset by the developer or even re-generate entirely new visualizations. It is difficult to meet the needs of "dynamic adjustment, instant feedback, and multi-dimensional relevance" in the exploration of complex problems in science (Andrienko *et al.*, 2020).

The emergence of interactive and immersive scientific visualization has completely changed this traditional paradigm. It has upgraded the visualization interface from a "static display window" to a "dynamic scientific laboratory," giving users the ability to actively manipulate data and explore real-time laws. Here, the user is no longer a passive bystander but becomes the dominant player in data exploration through interactive technologies (such as mouse drag and drop to rotate three-dimensional models, gestures scale time series, and voice commands filter subsets of data). Immersive technologies such as wearing a VR headset to "step inside" a molecule to observe the arrangement of atoms or superimposing virtual graphs onto a real experimental scene with the help of AR glasses can enable multi-dimensional perception beyond the confines of the screen. This "active exploration" model not only improves cognitive efficiency (the user can quickly focus on key areas), but it also expands cognitive boundaries (discovering hidden patterns and relationships that are difficult to detect in static views) and makes scientific visualization move from "data visualization" to "cognitive visualization."

The core value of interaction and immersion technology is that it restructures the process of scientific discovery through a depth of "human–machine–data" co-operation:

- Interaction breaks the "preset boundaries" of traditional visualizations, allowing users to dynamically adjust the visualization parameters according to their research objectives (e.g., modifying the time range, switching between variable combinations, adjusting rendering precision), and validate hypotheses and correct ideas with real-time feedback (e.g., dynamically updated charts and instantly highlighted association areas). For example, in the exploration of climatic analog data, scientists can observe the year-to-year variation of global temperature by dragging the timeline, clicking on a specific region to zoom in to see local climate anomalies, or even simulating future scenarios at different greenhouse gas concentrations through parameter tuning significantly speeds up the process of validating scientific hypotheses.
- Immersion breaks through the physical limitations of two-dimensional screens and through three-dimensional space layouts (such as mapping complex data fields to three-dimensional coordinates). Natural interactive modes (hand gestures and eye tracking) and multi-sensory interactions (visual, auditory, and tactile feedback) provide users with an immersive data exploration experience. For example, in medical image analysis, doctors can "walk" into the patient's three-dimensional organ model through VR equipment to visually observe the spatial relationship between tumors and surrounding tissues. In geological exploration, researchers can "dive" into the Earth's crust with a first-person perspective to track the extended path of fault lines and changes in rock structure. This immersive observation often inspires intuition and

reveals key details that are easily overlooked in traditional plane views (Ware, 2020; Keim *et al.*, 2021).

This chapter focuses on the core technologies, design methods, and application scenarios of interactive and immersive scientific visualization:

- The underlying logic of interactive technology — from basic mouse/keyboard interaction to advanced gesture recognition, voice control, and eye tracking, and how different interactive modes can be adapted to different scientific scenarios (e.g., high-precision interaction for precise operation, natural interaction for quick browsing);
- The path to implementation of immersion technology — from virtual reality (VR) three-dimensional enclosed spaces to augmented reality (AR), from three-dimensional graphic rendering to multisensory feedback fusion, and how to build a "data immersion field" by technical means;
- Co-design of interaction and immersion — how to balance degrees of freedom and visual clarity of interaction (to avoid confusion due to excessive manipulation), and how to align immersion experiences with scientific goals (to ensure that immersion serves cognitive deepening rather than mere entertainment);
- Typical application cases — from virtual laboratories in quantum physics to immersive Earth models simulating climate change, from interactive exploration of molecular structures to AR assistance for medical surgical planning, how interactive and immersive technologies drive cognitive revolutions in different areas of science;
- Interactive and immersive scientific visualization is not only a technological upgrade but also an innovation of scientific cognitive paradigm. Rather, it is a touchable, actionable, immersive "cognitive companion," which opens a new chapter of more imaginative and creative human exploration (Liu, *et al.*, 2008).

5.2. Interaction Design Principles

Interaction design is the core driving force of scientific visualization from "static display" to "dynamic exploration." In traditional scientific visualization, users are often "passive recipients" — data are presented through fixed charts, graphs, or animations, and users can only observe preset results. In interactive and immersive scientific visualization, users become "active explorers," dynamically adjusting the perspective, scope, and expression of visualizations through direct dialogue with data through operations (such as clicking, dragging, and zooming), inputs (such as voice and gestures), or immersive experiences (such as roaming in virtual reality), so as to discover patterns, associations, and anomalies in the data more deeply (Wimberly, 2023).

The core goal of interaction design is to build a set of rules that conform to the user's operational instinct, support efficient exploration, and adapt to the logic of scientific analysis so that users can establish a close connection with the data through intuitive interactive behaviors (such as direct manipulation of data, instant feedback, and gradual in-depth exploration). After a long period of practice and theoretical summary, interaction design in the field of scientific visualization usually follows three core principles: direct manipulation, immediate feedback, and progressive exploration. These three principles jointly support an efficient, intuitive, and scientifically valuable interactive experience from three dimensions: operation mode, response logic, and cognitive path. This section analyzes these three principles in detail and explains their specific applications in scientific visualization scenarios (Shneiderman, 1997; Heer & Shneiderman, 2012).

5.2.1. *Direct Operation: Make Data Manipulation as Intuitive as "Manipulating Physical Objects"*

Core idea: Users interact with data objects (e.g., data points, graphic elements, and view areas) through direct, physically intuitive interactive actions (e.g., grab, move, and adjust) rather than through complex menu commands or indirect parameter settings. The core of direct operation is to transform abstract data operations into familiar daily actions, reduce the learning cost of operations, and enable users to focus on the data itself rather than the interaction process (Javed & Elmqvist, 2016).

Embodied in scientific visualization:

- **Direct manipulation of data objects:** Users can directly select, drag, or adjust visualization elements related to the data. For example, in the scatter plot visualization, users can directly drag a data point with the mouse, and the system will update its corresponding data value according to the new position of the data point in real time (e.g., after moving the point from the original position, the system adjusts the value of a variable of the sample according to the preset rules and updates other related views to display the corresponding changes). In the 3D molecular structure visualization, users can directly grasp the atoms in the molecule and change their spatial position by dragging them to simulate the dynamic change process of the molecular structure and intuitively observe the impact of distance and angle changes between atoms on the overall shape of the molecule.

- **Direct control over visual views:** Users can adjust the presentation of their views through direct actions, making it as natural as manipulating a real object. For example, use the mouse wheel to directly zoom in on the visual graphic, zoom in on a specific area to observe the details (such as zooming in on the meteorological data map of a certain geographical area, to see the fine distribution of local meteorological elements), and zoom out to view the overall overview, panning the view through mouse drag to change the viewing angle (such as panning to view data from different geographical locations in the map visualization). It can also directly rotate the 3D visualization model (such as the 3D meteorological model of the rotating earth, observe the distribution of global meteorological elements from different angles) and explore data features from multiple directions (Ware, 2020).
- **Intuitive Operation Mapping Relationship:** There is a clear and intuitive mapping relationship between interactive actions and data changes, allowing users to easily understand what kind of data results their actions will bring. For example, when a user directly draws a rectangular area on a color spot map representing the temperature distribution, the system can immediately understand that the user's intention is to filter or focus on the temperature data in the rectangular area and then highlight the area or pop up the statistical information of the temperature data in the area. If a user directly drags a time point marker on a time series line chart, the system can intuitively adjust the time range according to the drag operation to re-display the data change trend within that time range (Sacha *et al.*, 2020).

5.2.2 *Instant Feedback: Seamless Connection Between Work and Response*

Core idea: When a user performs any interaction, the system must give clear, intuitive feedback in a very short time (ideally less than 100 milliseconds), informing the user that the action has been successfully received and that the system status has been updated accordingly. Instant feedback is the key to maintaining user fluidity and avoiding confusion, allowing users to confirm in real time that their actions have produced the desired results during operation, so that they can continue to explore data with more confidence (Nielsen, 1994).

Embodied in scientific visualization:

- **Instant Presentation of Visual Feedback:** This is the most common form of feedback, where a user's action immediately causes a visual change in the visual element. For example, when a user clicks on a data point, the data point instantly changes color (e.g., from a normal color to a highlighted bright color), size (e.g., enlarged to highlight), or shape (e.g., from a circle to a square), allowing the user to clearly see the selected target. When the user drags the slider to adjust a certain data parameter (such as temperature threshold), the visual graph (such as the temperature distribution chart) will be updated and displayed according to the new parameter value in real time, and the user can immediately observe the impact of parameter changes on the data presentation (such as the color change of the area above the new threshold). If the user performs data filtering, the filtered data will disappear immediately, and the remaining data will be quickly rearranged, making the whole process smooth and intuitive.
- **Assisted reinforcement of auditory and haptic feedback (optional but effective):** In some specific scenarios, in addition to visual feedback, auditory and haptic feedback can further enhance the effect of immediate feedback. For example, in the immersive environment of virtual reality scientific visualization, when a user successfully completes a key data interaction operation (such as accurately selecting an important data node hidden in a complex structure), the system can play a short, crisp prompt sound (such as a "ding" sound) to allow the user to audibly confirm the success of the operation. When users interact with devices such as controllers, if the operation triggers important events (such as adjusting to a critical time node or parameter state),

the handle can generate slight vibration feedback to enhance the user's perception of operation tactually. However, these feedback methods should be used carefully to avoid unduly interfering with the user's primary visual experience (Slater & Sanchez-Vives, 2016).

- **Clarity and accuracy of feedback information:** The feedback content must accurately convey the results of the operation and relevant information, without ambiguity. For example, when a user adjusts the timeline range, in addition to visual graphical updates, the currently selected time range (such as "January 2020–June 2020") can be displayed in real time at a fixed location in the interface (such as next to the timeline), allowing the user to clearly know the currently observed data period. If the user's actions prevent some data from being displayed (such as not having eligible data due to too strict filtering conditions), the system should provide a clear prompt message (such as "There is no eligible data under the current filter conditions, please adjust the filter parameters") instead of making users wonder why they are not seeing any data (Heer, 2021; Tufte, 2022).

5.2.3. *Progressive Exploration: Guide Users to Gradually Deepen Their Data Cognition*

Core idea: Considering that scientific data often have a high level of complexity and rich information, it is difficult for users to understand all the data details and internal laws at once. Therefore, interaction design should follow the principle of progressive exploration, provide users with a path and way to gradually explore from macro to micro, from simple to complex, from overall overview to local details, to help users gradually discover patterns, correlations, and anomalies in the data, and avoid cognitive overload caused by massive complex data (Pirolli & Card, 2020).

Embodied in scientific visualization:

- **Gradual in-depth from macro-overview to local details:** The system first presents users with an overall macro view of the data, allowing them to have a preliminary and comprehensive understanding of the overall distribution, general scope, and main characteristics of the data. For example, in a visual map of global meteorological data, the overall distribution of meteorological elements (such as temperature and precipitation) in various regions of the world is displayed in the initial state, allowing users to quickly understand the general pattern of global meteorology. When a user becomes interested in a specific area (such as a country or region), the system can quickly zoom in on the area and display it through simple interactions such as clicking or checking on the area showing detailed meteorological data of local areas (e.g., finer temperature gradients or precipitation distribution details) to achieve a smooth transition from macro to micro (MacEachren, 2021).

- **Step-by-step guidance from simple to complex analysis:** Interaction design should guide users to deeper data analysis based on their cognitive processes. In the initial stage, users are provided with some basic and simple interactive functions and views (such as basic data filtering and common visual chart types) so that users can quickly get started, conduct preliminary exploration, and simple analysis of data (such as filtering out data in a certain period of time and observing its basic change trend). As users gradually deepen their understanding of data, the system can gradually unlock or recommend more complex interactive functions and analysis methods (such as multivariate association analysis, advanced statistical calculations, and data mining algorithm applications) to guide users to discover more hidden patterns and associations in the data. For example, in a social science research data visualization containing multiple variables (such as population, economic indicators, and environmental parameters), users can initially understand the basic distribution of each variable through simple univariate screening and bar chart display. After that, the system can guide users to perform multivariate association analysis, explore the interrelationships between variables through interactive operations (such as selecting multiple variables and generating scatter plot matrices), and gradually dig deeper into the complex laws behind the data (Sacha *et al.*, 2020).
- **Interactive support based on user cognitive rhythm:** Respects the user's cognitive rhythm and personal learning curve, allowing users to explore data at their own pace and in their own way. The system should not force users to accept a large amount of complex information or perform difficult interactive operations at one time but provide flexible interaction methods and step-by-step guidance prompts. For example, for users who are new to a complex scientific visualization system, the system can provide a beginner guide mode with step-by-step prompts such as "Step 1: Observe the distribution of the overall data first, you

can zoom and drag the view to see different areas; Step 2: If you want to learn more about an area, you can click on the area to enlarge ......") to help users gradually familiarize themselves with the interactive functions and data exploration methods of the system. For experienced users, the system can intelligently recommend the next interactive operation or analysis direction that may be of interest based on the user's historical operation records and preferences, assisting users in more efficient data exploration, but always allowing users to adjust and explore independently according to their own wishes (Heer & Shneiderman, 2012).

The three principles synergistic support scientific interactive experience

The three interaction design principles of direct operation, instant feedback, and progressive exploration jointly build an efficient, intuitive, and interactive exploration experience in scientific visualization from different dimensions. Direct operation makes data operations as natural and intuitive as manipulating physical objects, lowering the threshold for users to interact with data, allowing users to focus on the data itself. Instant feedback ensures that every action of the user can receive a timely and clear response, maintain the smoothness and certainty of the operation process, and enhance the user's trust in the interactive system. Progressive exploration fully considers the complexity of scientific data and the progressiveness of user cognition and provides users with a step-by-step in-depth path from macro to micro, from simple to complex, helping users gradually discover the scientific laws and values in data without being overwhelmed by massive information.

These three principles are interrelated and complementary — direct actions provide the foundational way for users to interact with data, with instant feedback ensuring the effectiveness and fluidity of this interaction, while progressive exploration guides users to explore data in depth through direct actions and instant feedback, achieving cognitive sublimation from superficial observation to deep understanding. In the following chapters, we further explore how to combine specific interaction technologies (such as gesture recognition, voice interaction, and multi-view linkage) and immersive environments (such as virtual reality and augmented reality) based on these three principles to achieve the specific implementation of these interaction design principles in scientific visualization, so as to provide stronger and friendlier interactive visualization support for scientific research, science education, and science communication.

5.3. The Core Characteristics and Technological Foundations of Visual Immersive Presentations

In the system of interactive and immersive scientific visualization, "visual immersion" is one of the key dimensions to build the user's deeply engaged experience. It transforms abstract scientific data into a "perceptible scene" with spatial depth, dynamic continuity, and emotional resonance through a multisensory collaborative visual design strategy. This allows users to not only "see" the data but also "be" in the scientific phenomena the data describes, triggering embodied cognition — a more intuitive scientific understanding through the body's interaction with the environment. This section analyzes how visual immersion becomes an important extension of scientific visualization from three aspects: technical principles, design strategies, and typical cases.

5.3.1. *Core Features and Technical Foundations of Visual Immersive Presentation*

The essence of visual immersive presentation is to break the "observer perspective limitation" of traditional two-dimensional charts or static three-dimensional models through the spatial reconstruction of visual elements, dynamic narrative, and multimodal fusion, creating an experience where the user is immersed as a participant in the scientific scene (Stanney *et al.*, 2021). Its core characteristics can be summarized as follows:

1. **Spatial depth and proportional reduction:** This is through 3D stereo modeling or panoramic view area expansion (such as 180°/360° viewing), reducing the actual spatial structure of scientific objects (such as the stacking relationship of molecular orbits and the stereoscopic distribution of geological faults) and strictly following the real proportions (such as celestial dimensions contrast and micron-scale accuracy of organelles), which helps users establish "physical world-class" spatial intuition.

2. **Dynamic continuous and process visualization:** This captures the temporal evolution of scientific phenomena (such as the air flow cycle of a typhoon). Through key frame interpolation, fluid simulation, or particle system, discrete data can be transformed into continuous dynamic pictures (such as blood flow track in the blood vessel and slow motion playback of galaxy collision), revealing the hidden dynamic laws.
3. **Emotional empathy and enhanced ambience:** This is through light and shadow atmospheres (such as the gradual transition of deep-sea blue and purple). The color transition of the aurora flow, material details (such as the rough particles on the rock surface and the translucent texture of biological tissue), and sound effects (such as water flow and electromagnetic field buzzing) enhance the authenticity and emotional appeal of the scene and inspire the user's curiosity and exploration of scientific phenomena.

Its technical foundation relies on three types of tools:

- **3D graphics engines (e.g., Unity3D and Unreal Engine):** These support real-time rendering of high-precision models and dynamic light and shadow and enable interactive perspective switching (e.g., users freely rotate molecular structures and zoom in on nebulae details).
- **Panoramic/VR display devices (e.g., CAVE Immersive Projection System and Oculus Rift headset):** These provide a 1:1 ratio "first-person perspective" experience (such as standing inside a cell to observe mitochondrial activity) through multi-projection surface patchwork or headset field coverage.
- **Data mapping algorithms (e.g., Volume Rendering and Streamline Visualization):** These combine scientific data (e.g., temperature field, pressure distribution, and molecular orbital energy levels) into color gradients, transparency, or particle density, visualizing invisible physical quantities (such as magnetic field strength and chemical reaction rate).

5.3.2. *Four Strategies for Visually Immersive Design*

To achieve effective immersion of scientific data, design needs to balance "accuracy" and "perception," with a focus on the following strategies:

Strategy 1: Perspective Narration — From "God's Perspective" To "Participant's Perspective"

While traditional scientific visualization often takes a "big picture" perspective (e.g., map-based climate data distribution), immersive design puts more emphasis on "user engagement." For example:

- In meteorological visualization, users can "form" as water droplets in the atmosphere, follow the air flow up to the troposphere, and observe the process of condensation of water vapor into clouds (perspective moving with the air flow, surrounded by rolling clouds and lightning).
- In the astronomical visualization, users can "stand on the surface of Mars" and look up at the Earth as a blue-green planet hanging in an orange-red sky (by adjusting the headset field to feel the distance and relative size of the Earth and Mars). This "first-person/restricted perspective" design translates abstract data (e.g., air pressure values and orbital parameters) into spatial orientation and motion perception (e.g., "I am crossing a thunderstorm zone" and "The Earth looks smaller than I thought") by simulating the user's actual position in a scientific scene.

Strategy 2: Hierarchical Information Presentation — Visual Guidance That Focuses on Key Data

Information overload in immersive environments can easily lead to cognitive load, so we need to highlight core scientific variables through visual hierarchy design. Common methods include the following:

- **Foresight-background separation:** This includes positioning key data (e.g., location of a typhoon eye and key mutation sites in cancer cells) in a sharp contrast color (e.g., fluorescent red) or dynamic effect (i.e., pulsed halo). Secondary backgrounds (e.g., ocean surface ripples, normal tissue profiles) are treated with low transparency or blurring.
- **Interactive focus:** This includes dynamic switching of the scale of observation (e.g., scaling from the group of galaxies to a single star) through gestures (e.g., pinching), gazing (e.g., gazing at an area for 3 seconds), or clicks.
- **Symbolic marking:** This adds simplified visual signs (such as arrows for current directions and color blocks for thermal ranges) to avoid

text-only tags that spoil the immersion (such as dynamic light strips instead of "magnetic field strength + 5T").

Strategy 3: Multisensory Synergy — Enhancing the Connection Between Vision and Touch/Hearing

Visual immersion does not exist in isolation but forms a "sensory cluster effect" through synergy with tactile feedback (such as the vibration of the handle simulates seismic wave strength) and auditory cues (such as changes in the frequency of blood flow correspond to blood pressure levels). For example:

- In geological visualization, when the user observes plate collisions, the headset picture synchronizes the dynamic process of the mountain rise, the handle vibrates with the increased pressure of the squeeze (strength corresponds to pressure value), and the background sound changes gradually from a low dull sound to a sharp rock friction sound.
- In biological visualization, when cardiac muscle cells are observed to contract, the muscle fibers in the picture shorten and glow synchronously (visual), the headphones emit a thump of heartbeat (the auditory frequency is in sync with the contraction), and the user can even "touch" the elastic changes of the cell membrane through the tactile glove.

Strategy 4: Articulate Interaction — Deepen Understanding Through Body Movements

Active participation of the user's body reinforces the understanding that data is the environment. For example:

- In fluid mechanics visualization, the user simulates "applying external forces" (such as stirring liquids) by gesture, observing the formation and dissipation of vortices (the movement trajectory of the hand directly affects flow field changes).
- In molecular biology visualization, users "reach out" to grab DNA double helix structures and observe the dynamic matching of base pairs by rotating and stretching (hand movements and molecular bonds can change in real-time feedback).
- In astrophysics visualization, the user "walks" between the orbits of the solar system planets, adjusts the speed of time flow by the pace

(the faster the walk, the faster the planets turn), and intuitively feels the relationship between gravity and distance.

5.3.3. *Typical Case: Practical Verification of Visual Immersion Scientific Visualization*

Case 1: NASA "Mars 2030" VR experience (Astronomy/Geosciences)
The project uses Unreal Engine to build a high-precision three-dimensional model of the Martian surface (based on real terrain data from the Curiosity rover). After wearing a VR headset, users can "stand" on the edge of Mars' Jezero crater, observe the granular details of red soil, the outline of a distant crater, and even crouch down and "touch" the weather layer on the rock surface with a virtual hand. In a dynamic view, users can follow the Martian sunset to observe the sky changing from blue to pink to purple (based on a simulation of light scattering from the Martian air) and collect "virtual rock samples" through interaction with the handle (triggering a geo-analysis window). Through immersion of being on Mars, the design helps the public and researchers to more intuitively understand the geological evolution and habitability conditions of Mars.

Case 2: CERN "Particle Collider" immersive laboratory (particle physics)
The immersive system developed at CERN uses the CAVE projection environment (full projection of four walls and ground) to turn experimental data from the LHC (Large Hadron Collider) into a visualization of dynamic particle flows: The user stands in the middle of the "collision point," the surrounding projection presents the quark-gluon plasma produced after a proton beam collides at nearly the speed of light (in the form of a colorful energy cloud), and the particle trajectory is tracked in real time via a linear animation (color corresponds to energy intensity). When the user raises his hand, the system simulates a "magnetic field intervention" and the particle flow path deflects (interaction feedback delay < 50 ms). The design allows physicists to observe the instantaneous behavior of microscopic particles "in the realm" of experiments, helping validate the standard model theory.

Case 3: Ocean Acidification Simulator (Environmental Science)
In response to the problem of ocean acidification in the global climate change, a team has developed an immersive experience based on Oculus Quest 2: The user "dives" into the future (2100) coral reef waters, the originally colorful corals were observed to bleach gradually due to a decrease in pH (from bright orange-red to gray-white visually), and the surrounding seawater presented a muddy mist sensation due to increased dissolved carbon dioxide (through fog effect particle simulation). When a user swims, the silence of the decrease in marine life after the death of coral worms (compared to the chirping of fish in normal waters) is heard in the ear, and a handheld sensor "measures" the pH of water at different depths (the values are displayed in real time on a virtual panel). The design reinforces public empathy and scientific understanding of the impact of carbon emissions through a sense of "personal experience of ocean decline."

5.3.4. *Challenges and the Way Ahead*

Although visually immersive presentations have significantly increased participation in scientific visualization, there are challenges to its growth:

- **Technical hurdles:** High-precision modeling and real-time rendering require high hardware computing power (such as 8K resolution VR needs GPU support), which may limit the popularity.
- **Risk of cognitive bias:** Overemphasis on visual shock may cause users to ignore the rigor of the data itself (e.g., exaggerating the intensity of phenomena in pursuit of "shock effects").
- **Universal adaptation:** Users of different age groups/cognitive levels vary greatly in their needs for immersion depth (e.g., children are more suitable for simple dynamic narratives, and researchers need more precise data labeling).

In the future, with the development of lightweight XR devices (such as AR glasses), AI-driven adaptive rendering (which dynamically adjusts the level of detail based on the user's focus of attention), and multimodal fusion technologies (such as tactile feedback combined with olfactory simulation), visual-immersive scientific visualization will further evolve in the direction of "more real, more accurate and more inclusive" and become a key bridge between scientific data and public perception.

Visual immersive presentation is an important part of interactive and immersive scientific visualization. It transforms scientific data into "accessible, perceptible, and interactive" scenarios through spatial reconstruction, dynamic narrative, and multi-sensory collaboration. Designers need to find a balance between scientific accuracy, visual appeal, and cognitive friendliness to achieve the goal of making science visible, touchable, and remembered.

Part 3

Practice: Creative Landing from Project to Work

Chapter 6

Process and Methodology of Scientific Visualization Projects

6.1. Introduction: The "Project Management Compass" of Scientific Visualization

Scientific visualization is not a technical exercise in isolation. Rather, it is a project that moves from "problem in science" to "visual insight" — starting with the need to understand complex data. "How can I show the dynamic process of protein folding?" Finally, users — scientists, engineers or the general public — receive intuitive understanding and decision support through a visual interface. However, this process is far from a simple "data input–graphics output," but rather a complex collaboration involving requirements analysis, data adaptation, technology selection, design iterations, evaluation and optimization, each of which may affect the scientific and effectiveness of the final visualization (Kosara, 2021).

The success of scientific visualization depends on a clear process framework and scientific methodology. Just as building a bridge requires a complete process from geological survey, structural design, and construction supervision, scientific visualization also requires a systematic set of process specifications: From the "why" (scientific objectives and user needs), define "what to do" (data types and visual tasks), plan "how to do" (technical tools and design strategies), and verify "how to do" (impact assessment and continuous optimization). This process not only ensures that visualizations are always focused on the core problem in science but also avoids common misconceptions (such as cognitive burden caused by overdesign, technical skills deviating from actual needs) through standardized steps, and finally achieves the precise match of "data–technology–customer" (Munzner, 2022).

At the same time, as an interdisciplinary field of science and technology, the methodology of scientific visualization needs to follow the logic of data science (such as data cleaning, feature extraction, and multivariate association analysis), rely on computer graphics implementation techniques (such as 3D modeling, rendering optimization, and interactive development), and also need to incorporate user cognitive laws (such as visual perception preferences, and information processing efficiency) into cognitive psychology. Together, these methodologies form a "design guide" for visualization projects, guiding developers from "technical implementation" to "cognitive services," for example, optimize graphic layout through the visual organization principles of Gestalt psychology, use the theory of human–computer interaction to design intuitive operational logic, and balance information density and understanding efficiency based on cognitive load theory (Lam *et al.*, 2020).

This chapter will systematically analyze the whole process and core methodology of scientific visualization projects, focusing on the following:

- **Phases and critical tasks of the project process:** These include requirements research (deep communication with scientists/users to clarify scientific objectives and visualize uses), data preparation (gathering, cleaning, and structured processing to ensure data quality and availability), task definition (breaking down specific visualization objectives, such as comparison, distribution analysis, and relational reasoning), and technology selection (matching tools and algorithms according to data characteristics to task requirements). From design and development (visual coding, interaction logic, and interface prototyping) to testing and evaluation (user feedback collection, cognitive performance validation, and iterative optimization), the "life cycle" of a visual project is presented completely.
- **Practical application of core methodology:** This includes how to select the optimal visual mapping based on the task–data–coding theory (e.g., location/length encoding of quantitative data and classification data by color/shape), how to lower the threshold of user understanding following cognitive design principles (e.g., data ink ratio optimization, visual metaphor design), how to increase exploration

depth through interactive techniques (e.g., dynamic filtering and multi-view interaction), and how to evaluate the science and usability of visualizations (e.g., task completion time measurement, error rate statistics, user satisfaction survey).

- **Key points of interdisciplinary collaboration:** Analyzing common collaborative scenarios in scientific visualization programs (e.g., scientists providing data against research goals), designers optimize visual presentation, engineers implement technical functions and summarize ways to communicate across disciplines (such as accurately describing requirements in scientific terms, rapidly validating concepts with visual prototypes, and bridging cognitive differences through iterative feedback) to ensure teams work together effectively.

The process and methodology of the scientific visualization project essentially constitute the practical wisdom that integrates the "logic of scientific discovery" and "design of user experience." It is not only a set of operational guidelines but also a key support for ensuring that visual projects move from "technical realization" to "scientific value" — through a standardized process and a scientific approach. Every visual work can accurately convey the scientific story behind the data and become a "visual engine" that drives scientific research, engineering practice, and public awareness.

6.2. Requirements Definition: Precise Anchoring from Scientific Problems to Visual Goals

The essence of a science visualization project is to transform scientific data into intuitive, interactive visual representations to support scientific discovery, decision-making analysis, or knowledge dissemination. However, this process is not a simple combination of "data + graphics" but begins with a deep understanding of scientific problems and ends with a precise definition of visualization goals. The requirements definition stage is the starting point and cornerstone of the entire project — it determines the subsequent data selection, visual form design, technical implementation path, and even the scientific value and practical significance of the final work. If the requirements are vaguely defined, the follow-up work may fall into the misunderstanding of "visualization for the sake of visualization," resulting in results deviating from scientific goals (such as excessive pursuit of visual effects and losing key data information) or failing to meet the real needs of users (such as researchers needing analysis tools that become popular science displays for the public).

This section will focus on "requirements definition" to systematically analyze how to extract clear visualization goals from complex scientific problems, clarify the core constraints and success criteria of the project, and ensure that the visualization solution is closely aligned with the scientific problem through a structured requirements analysis framework.

6.2.1. *The Core of Requirements Definition: The Mapping of Scientific Problems and Visual Goals*

The essence of the requirements definition is to answer three key questions: (1) "What scientific problem is to be solved?," (2) "What core message is conveyed through visualization?," and (3) "Who will it serve in the end and what effect will it achieve?" These three questions correspond to the nature of the scientific problem, the core of the visualization goal, and the user and value orientation of the project, which together constitute the "iron triangle" of requirements definition.

1. The Nature of Scientific Questions: Clarifying "Why Visualization"
Scientific visualization does not exist to "show data," but to solve specific scientific problems — whether it may be to test a hypothesis (e.g., "does climate change cause a species to shift northward?"), discover hidden patterns (e.g., "anomalous associations in multivariate sensor data"), optimize decisions (e.g., "adjust industrial production parameters based on real-time monitoring data"), or disseminate scientific knowledge (e.g., "explain the fundamentals of quantum computing to the public"). Therefore, the first step in defining requirements is to communicate deeply with scientists, engineers, or domain experts to clarify the scientific background, research objectives, and core issues to be solved.

Practical points:

- **Clarify the scientific motivation through "problem traceability":** For instance, if the project originates from a research gap in a paper (e.g., "existing models cannot explain the spatial heterogeneity of an ecological phenomenon"), it is necessary to trace the background of the problem (e.g., limitations of observational data, hypothetical flaws of the model) to determine the information that needs to be supplemented by visualization (e.g., spatial distribution details, multivariate interactions).

- **Distinguish between "scientific objectives" and "visualization objectives":** Scientific objectives may be "quantifying the impact weight of a parameter" (supported by data analysis), while visualization objectives may be "visualizing the correlation pattern between the parameter and other variables" (through graph-assisted analysis); avoid mistaking data analysis tasks (such as statistical testing) for visualization needs (e.g., "just draw a line chart").

2. The Core of The Visual Goal: Define "What Message to Convey"

Based on scientific questions, it is necessary to further clarify the core types of information that need to be conveyed by visualization (such as trends, distributions, associations, and anomalies) and key cognitive needs (such as "users need to compare data differences in different regions" and "need to observe the dynamic changes of parameters over time"). Visualization goals should be specific, measurable, and directly serve the solution of scientific problems.

Common types of visualization goals:

- **Descriptive objectives:** Show the objective status or distribution characteristics of the data (such as "the annual average temperature change of a certain region in the past ten years" and "the pH value range of different soil types") to help users establish a basic understanding of the data.
- **Analytical objectives:** Reveal patterns, correlations, or anomalies in the data (e.g., "the negative correlation between temperature and humidity is only significant in summer" or "periodic mutations in a sensor data") to support scientific hypothesis verification or problem diagnosis.
- **Exploratory objectives:** Assist users in discovering unknown patterns, or potential hypotheses (e.g., "the hidden clustering mode in multivariate sensor data" and "new correlation clues under user interaction exploration") are often used for data mining in the early stage of scientific research.
- **Explanatory objectives:** Convey scientific concepts or conclusions to a non-specialist audience (e.g., "explain how vaccines activate the immune system with dynamic graphics" or "illustrate the effects of

new energy policies through comparison charts"), emphasizing intuitiveness and accessibility.
- **Decision-making objectives:** Provide a basis for action (e.g., "recommend equipment maintenance time based on real-time monitoring data" or "assist in policy development through risk visualization"), highlighting key indicators and thresholds.

3. **User and value orientation: Clarify "who to serve and what effect to achieve"**

Visualization end users (e.g., researchers, engineers, policymakers, and the public) determine their needs — researchers need high-precision, interactive analysis tools, and the public is more concerned about intuitiveness and storytelling; decision makers need a clear presentation of key indicators, while students may need step-by-step learning guidance. Therefore, the requirements definition needs to clarify the characteristics of the target user group (expertise, use scenarios, technical proficiency) and define the success criteria of the project (e.g., "help researchers save 50% of data exploration time" and "improve public understanding of scientific concepts by 30%").

Practical points:
- **User profile analysis:** Through interviews, questionnaires, or observations, understand the specific needs of users (e.g., "doctors need to view the correlation changes of multiple physiological indicators of patients in real time" and "teachers need visual tools to assist in explaining the energy flow of ecosystems"), usage scenarios (such as laboratory analysis, classroom demonstrations, and mobile quick viewing), and pain points (e.g., "existing tools are complicated to operate" and "data display is not intuitive").
- **Value alignment:** Ensure that visualization goals are aligned with actual user needs (e.g., climate change visualizations designed for policymakers need to highlight risk levels and policy intervention windows in key regions rather than just showing global average temperature trends).
- **Definition of success indicators:** Quantitatively or qualitatively describe the success of the project (e.g., "users can find a subset of target data within 3 minutes," "number of times visualization results are cited," and "cognitive clarity score based on user feedback").

6.2.2. *Structured Process of Requirements Definition: From Vague Requirements to Clear Solutions*

Requirements definition is not a one-time "pat on the head" decision, but an iterative, structured exploration process that usually includes the following five key steps:

Step 1: Scientific background research and question clarification
Objective: To understand the scientific origins, research status, and core issues to be solved of the project.

Practical method:

- Conduct in-depth interviews with subject matter experts (scientists, engineers, and business leaders) to understand the context of the project (e.g., "Why study this issue? and "What data and analysis methods are there now?"), core scientific questions (such as "What is the hypothesis that needs to be tested?," "What is the pattern that is expected to be discovered?"), and existing challenges (e.g., "data dimensions are too high to analyze" and "existing visualization tools cannot meet the needs").
- Consult relevant literature (papers, reports, and industry standards) to identify the need for visualization of similar problems in the field (e.g., "What charts are commonly used in meteorology to show temperature field distributions?" and "How can spatial-temporal patterns of gene expression be visualized in biology?"), and avoid reinventing wheels or deviating from domain conventions.
- Sort out existing data resources (such as sensor data, experimental records, and simulation results) and preliminary analysis conclusions (e.g., "preliminary findings of seasonal correlation between temperature and humidity"), and clarify the availability and limitations of data (e.g., "there is a 10% missing value in the data" and "the time resolution is only in the hour level").

Step 2: Extract core information requirements
Objective: Extract the key information types and user operation needs that should be conveyed by visualization from scientific problems.

Practical method:

- Through the "information pyramid" tool, the scientific problem is decomposed layer by layer: The top layer is the core scientific goal (e.g., "explaining the cause of a phenomenon"), the middle layer is the key information that supports the goal (e.g., "the distribution, correlation, and dynamic change of related variables"), and the bottom layer is specific data indicators (e.g., "temperature, humidity, wind speed values and time series").
- Discuss with users "biggest concerns" (e.g., "I most want to know what conditions are parameter A and parameter B" and "users need to compare data differences in different regions in real time") and convert it into visualization requirements (e.g., "bivariate scatter plots are required to display the relationship between parameters A and B, and support filtering by region").
- Clarify the user's operational needs (such as "need to interact to filter data for a specific time period" or "need to dynamically adjust parameter observation changes"), which will directly affect the subsequent interaction design.

Step 3: User and Scene Definition
Objective: Clarify the target user groups, usage scenarios, and constraints of the visualization service.

Practical method:

- Draw a "user–scenario–demand" matrix: List all possible user types (such as researchers, engineers, students, and the public), describe typical usage scenarios for each user (such as "researchers analyze data in the laboratory," "students learn concepts in class," and "public view popular science information on mobile phones"), and correspond to their core needs (such as "high-precision analysis," "intuitive learning," and "quick conclusions").
- Identify key constraints (e.g., time constraints ("project to be delivered within 3 months"), technical constraints ("can only be deployed on the web and no local software can be installed"), data constraints ("sensitive data needs to be desensitized"), and display environment restrictions ("need to adapt to large screen projection or small mobile screen").
- Define the user's cognitive level and technical proficiency (e.g., "professional users can accept complex interactions and ordinary users need simple guidance"), which will affect the complexity of visual design and the choice of interaction methods.

Step 4: Visualize the materialization and prioritization of goals
Objective: Transform abstract visual requirements into concrete, actionable goal descriptions and prioritize each goal.

Practical method:

- Define each visualization goal with the "SMART principle" (specific, measurable, attainable, relevant, time-bound): For example, "show the monthly trend of PM2.5 concentration in a city over the past year" is more specific than "make a map of air quality"; "users can find data in the target area within 5 seconds" is more measurable than "easy for users to find data."
- Prioritize multiple visualization goals (e.g., use the Kano model to distinguish "essential requirements" (such as data accuracy), "desired requirements" (such as interaction convenience), and "exciting needs" (such as cool dynamic effects)) to ensure that limited resources are prioritized to meet core requirements (such as data accuracy and analysis depth for scientific research projects, and intuitiveness and ease of understanding for popular science projects).
- Identify key cognitive points that need to be reinforced for visualization (e.g., "users need to focus on the negative correlation between parameter A and parameter B" and "anomalous data points need to be highlighted") that will guide subsequent visual coding design (e.g., color shades to indicate correlation strength and outliers to be marked with special symbols).

Step 5: Write and confirm the requirements document
Objective: Organize all of the above analysis results into a structured requirements document and confirm with all stakeholders (scientists, user representatives, and technical teams) to ensure a consistent understanding of the requirements.

Document content typically includes the following:

- project background and scientific issues (briefly describe the origin of the project, research objectives and core issues to be solved);
- core visualization objectives (prioritize key messages that need to be conveyed and corresponding user needs);
- target user groups and usage scenarios (describe in detail user characteristics, typical usage scenarios, and core requirements);
- data requirements (clarify the type, variables, time frame, and quality requirements of the data required);

- constraints (limitations on technology, time, data security, and display environment);
- success criteria (quantitative or qualitative project success indicators);
- initial visualization scenario (optional, e.g., "plan to use an interactive line chart to show trends and heat maps to show distributions") to provide direction for subsequent design.

6.2.3. *Common Misunderstandings and Avoidance Strategies of Requirements Definition*

In practice, the following misunderstandings often cause subsequent projects to deviate from the track in the requirements definition stage, and special attention should be paid to avoidance:

- **Myth 1. Confusing "what the user wants" with "what the user wants":** Users may offer to "make a beautiful dynamic graph" or "show all the data" However, this may not be the best solution to scientific problems. Through in-depth communication, it is necessary to tap into the real needs behind users (such as "beautiful dynamic maps" may be to attract public attention and "show all data" may be worried about missing key information) and guide them to focus on the core goals (Sedlmair *et al.*, 2021).
- **Myth 2. Ignoring the actual limitations of data:** Requirements that are defined without sufficient consideration of data integrity (e.g., a large number of missing values), accuracy (e.g., sensor margin of error), or timeliness (e.g., data updated only to the last month) can result in unreliable or unachievable visualizations. It is necessary to confirm the data quality with the data provider at an early stage and specify the data processing requirements in the requirements (e.g., "missing values need to be labeled" and "only use calibrated sensor data" (Wimberly, 2023).
- **Myth 3. Excessive pursuit of technological advancement and neglect practicality:** For example, forcing complex virtual reality (VR) interactions to present simple trend data not only increases development costs but can also make it difficult for users to operate. It is necessary to choose the appropriate technology according to user needs and scenarios, and prioritize the achievement of visualization goals rather than simply pursuing technical highlights (Hansen, *et al.*, 2014).

- **Myth 4. Requirements documents are vague or missing:** The needs of verbal communication are prone to subsequent differences due to differences in understanding (e.g., "interactive screening" specifically refers to filtering by time, space, or variables?). All requirements need to be clearly recorded through structured documents and confirmed with stakeholders through review meetings to ensure that "one document manages the whole process" (Munzner, 2022).

6.2.4. *Requirements Definition: The "Compass" of Scientific Visualization Projects*

The requirements definition stage is the "compass" of scientific visualization projects, which transforms vague scientific problems into clear visualization goals and abstract user requirements into concrete design constraints through structured analysis and communication. Only when the requirements are accurately defined can the subsequent data selection, visual form design, interactive logic implementation and technology selection be targeted, and the final visual works that meet scientific needs and have practical value can be produced. In the following chapters, we will further explore the whole-process methodology of data adaptation, visual form selection, interaction design implementation, and project implementation based on clear requirements definition.

6.3. Data Adaptation: Precise Transformation from Raw Data to Visual Driving Force

The requirements definition clarifies "what scientific problem to solve" and "what core information needs to be conveyed," while data adaptation is the key intermediate link in transforming raw data into supporting visualization goals — it determines whether the data are "usable, easy to use, and sufficient," which directly affects the accuracy, interactivity and scientific value of visualization. Scientific data are often multi-sourced (such as sensor networks, satellite remote sensing, and experimental recordings), heterogeneous (such as numerical, textual, and spatiotemporal data) and complex (such as high-dimensional, unstructured, and noisy), and if they are directly used for visualization without adaptation, it may lead to problems such as "data cannot drive graphics" (such as missing key variables), "graphics mislead cognition" (such as wrong time alignment), and "interaction failure" (such as data granularity does not match interaction requirements).

This section focuses on the core logic of data adaptation and analyzes how to filter, clean, transform, and structure the original data based on the visualization goals and user needs in the requirements definition, making it the "precision fuel" for adapting visual design. Specifically, data adaptation answers three key questions: (1) "Does the data cover the core variables needed for visualization?," (2) "Does the data meet the technical requirements for interaction and display?," and (3) "Are the data processed to support the reliability of scientific conclusions?" Around these problems, data adaptation usually includes the following four steps: data screening for requirement alignment, scientific and rigorous cleaning and preprocessing, structured transformation of adaptation visualization, and validation optimization of data-visualization mapping.

6.3.1. *Data Screening for Demand Alignment: Clarify "Which Data to Use"*

The requirements definition has clarified the core information that needs to be conveyed by visualization (such as "correlation pattern between temperature and humidity" and "spatial–temporal distribution of PM2.5 in a certain area"), so the first step in data adaptation is to screen out the key variables and data subsets that directly support this information from the original dataset to avoid "data redundancy interference" or "missing key information."

1. Demand Mapping of Core Variables

First, it is necessary to clarify the key variables (such as independent variables, dependent variables, and control variables in scientific problems) and auxiliary variables (such as time, spatial coordinates, sample IDs, and other metadata) that must be included according to the visualization goals. For example:

- If the visualization goal is "to analyze the relationship between crop yield and soil pH and precipitation," the key variables are "yield," "pH," and "precipitation," and the auxiliary variables may include "planting area coordinates" and "year."

- If the goal is "show the inter-annual trend of global temperature," the key variable is "temperature," and the secondary variables are "time (year/month)" and "geographic location (latitude and longitude)."

Practical method:

- Trace back to the visualization objectives in the requirements document and extract variables explicitly mentioned in it (e.g., "need to compare parameters A and B") or implicitly dependent variables (e.g., "show dynamic changes" necessarily require time variables).
- Confirm the scientific necessity of the variable with a scientist or domain expert (e.g., "does it have to be included?" and "does the lack of it affect the conclusion?") to avoid losing key information due to overscreening.
- Label the types of variables (e.g., continuous, discrete, subtype, spatiotemporal type) in preparation for subsequent cleaning and transformation.

2. Scoping of Data Subsets

Raw data may cover a wider range of time, space, or samples (e.g., global decadal meteorological data), but visualization goals may focus only on a specific subset (e.g., "PM2.5 data for eastern China from 2020 to 2023"). Therefore, it is necessary to define the time range, spatial area, sample category, etc. of the data according to the user scenarios and analysis goals defined in the requirements.

Common Qualifications:

- **Time range:** This includes "only display data from the past three years" and "filter out observation records in spring (March–May)."
- **Spatial scope:** This includes "focusing on the built-up area of a city" and "excluding sensor data from the marine area."
- **Sample categories:** These include "select only the experimental data of a certain type of crop (such as wheat)" and "eliminate samples marked with abnormal marks (such as sensor fault records)."
- **Variable value range:** This includes "only analyze data with a temperature range of 0–40°C," "filter out soil samples with pH between 5.5 and 7.5."

Practical method:

- Combine user needs in demand (e.g., "users need to observe changes in a specific area") and scientific problems (e.g., "a phenomenon only occurs under certain conditions"), and clarify the filtering rules for subsets.
- Quickly target subsets through data catalogs or metadata (e.g., timestamps, geolocation fields in data files).
- Record the logic of the filter criteria (e.g., "Time>=2020-01-01 AND Time<=2023-12-31 AND Longitude BETWEEN 116 AND 120 AND Latitude BETWEEN 39 AND 41") to ensure reproducibility and verification later.

6.3.2. *Scientific and Rigorous Cleaning and Preprocessing: Ensuring Data Availability*

The filtered subset of data may still have issues such as noise, missing, errors, or format incompatibilities, which can lead to graphical distortion (e.g., outliers deflecting trend lines), interaction failures (e.g., missing timestamps for dynamic playback), or unreliable scientific conclusions (e.g., incorrect data associations). Therefore, it is necessary to clean and process data based on scientific rigor to meet the basic quality requirements of visualization.

1. Missing Value Handling: Fill in or Reject

Missing values are a common problem (such as missing data at a certain point due to occasional sensor failure), and the treatment method should be selected according to the importance of the variable and the proportion of missing values:

- **Missing key variables (e.g., missing temperature data):** If the missing ratio is low (<5%), it can be filled by interpolation (e.g., linear interpolation and mean interpolation of neighboring points); if the missing proportion is high (>20%), consider removing this time period or sample ("incomplete data" should be marked in the visualization).
- **Missing non-critical variables (e.g., missing auxiliary metadata):** If it does not affect the core analysis (e.g., missing sample IDs but other variables are intact), the data can be retained and marked as missing; If it affects the data association (such as not being able to match the spatial–temporal information), it needs to be removed.

Practical method:

- Count the missing proportion of each variable (e.g., use Python's pandas library to calculate *df.isnull().mean()*), and determine the processing strategy proportionally.
- For time series data, preferential use of interpolation methods with time proximity (e.g., linear interpolation to maintain trend continuity); for spatial data, spatial interpolation can be used (e.g., Krigin interpolation estimates missing values based on adjacent points).
- Clearly indicate how missing data are handled in visualizations (e.g., "missing values are filled by linear interpolation") to avoid user misunderstandings.

2. Outlier Detection and Correction: Avoid "Noise Misdirection"

Outliers, such as extreme hot/cold temperatures recorded by sensor failures, can skew the visualization results (such as abrupt spikes in line charts). Outliers need to be identified and dealt with based on scientific background (e.g., normal value range of a parameter) and statistical methods (e.g., 3σ principle, box plot interquartile range).

Practical method:

- **Statistical detection:** Calculate the mean and standard deviation of variables (such as temperature mean of 20°C and standard deviation of 5°C), and mark the value that exceeds the mean value $\pm 3\sigma$ (i.e., outside 5–35°C) as suspected outliers, or define outliers by the upper and lower quartile range (IQR) of the box plot (e.g., values less than Q1–1.5IQR or greater than Q3+1.5IQR).
- **Scientific background verification:** Judge whether the outlier value is reasonable based on domain knowledge (e.g., although the extreme heat event in a certain area is beyond the conventional range, it may be a real climate phenomenon, and it needs to be retained and labeled).
- **Handling method:** If the outlier is caused by an error (such as sensor failure), it can be eliminated or corrected (such as replacing it with a nearby normal value); if it is a real but rare event (such as a sudden drop in temperature after a volcanic eruption), it should be retained and illustrated in the visualization with special symbols (such as red markers) or annotations.

3. Uniformity of Data Formats and Units: Ensuring "Comparability"

The original data may come from different devices or observatories, and there are problems, such as format differences (e.g., timestamp formats of

"2020/01/01" and "01-Jan-2020"), inconsistent units (e.g., mixing temperature units with °C and °F), and inconsistent encoding (e.g., "male/female" and "M/F" for gender variables), and need to be unified into a standardized format that can be handled by the visualization system.

Practical method:

- **Time format:** This is uniformly converted to standard format (e.g., "YYYY-MM-DD HH: MM: SS" of ISO 8601) for easy timeline interaction and dynamic playback.
- **Unit uniformity:** Convert the units of all variables to the same standard (e.g., temperature is unified as °C and length is unified as meters), and the units are clearly marked in the visualization.
- **Categorical variable encoding:** Converts text-based categorical variables to numerical or systematic A text label (e.g., "Soil Type: Sand/Clay/Loam" encoded as 1/2/3, or uniformly capitalized as "SAND/CLAY/LOAM") to avoid data association errors due to spelling differences.
- **Data type conversion:** Convert string-type values (such as "123" stored as text) to numeric types (such as integers or floating-point numbers) to ensure mathematical calculations (e.g., averaging and drawing continuous curves).

4. Data Standardization and Normalization (Optional but Critical)

If visualization involves multivariate comparison (e.g., temperature (0–40°C) and precipitation (0–200 mm)), excessive variable dimension differences may lead to graphical imbalances (e.g., precipitation values widen the axis and mask subtle changes in temperature). In this case, the variables need to be converted to the same comparable range by normalization (e.g., *Z*-score normalization) or normalization (e.g., Min–Max scaling to 0–1).

Practical method:

- **Standardization (*Z*-score):** This is suitable for scenarios where the distribution characteristics of the original data need to be preserved (such as the normal distribution hypothesis in scientific analysis), the formula is $X_{\text{std}} = (X - \mu)/\sigma$ (μ is the mean, σ is the standard deviation).
- **Normalization (Min–Max):** This is suitable for visualization scenarios (e.g., heat map color mapping) that need to map variables to a fixed

interval (e.g., [0,1] or [0,100]), the formula is $X_{norm} = (X - X_{min})/(X_{max} - X_{min})$.

- **Note:** Normalization/normalization is only for visualization, and raw data should be used if scientific calculations are required (e.g., regression analysis) to avoid distortion.

6.3.3. *Adapting to the Structured Transformation of Visualization: Building a Data Model That Can Drive Graphics*

The filtered and cleaned data needs to be further converted into a structured model that adapts to the visual design, so that it can be mapped directly to graphical elements (such as points, lines, and surfaces) or interactive logic (such as filter conditions and dynamic parameters). The essence of this process is to convert the semantics of scientific data (such as "the temperature at a certain time and place") into a visual syntax (such as "the temperature value at coordinates (x,y) corresponds to the shade of color").

1. Adjust Data Granularity: Match Interaction Needs
Interaction design may require different granularity of data (e.g., "overall trends need to be aggregated by year and detailed analysis needs to be displayed on a daily basis"), so you need to adjust the data granularity according to the interaction needs in the visualization objective:

- **Coarse-grained aggregation:** If users need to observe macro trends (e.g., "average annual changes in global temperature over the past decade"), they need to aggregate raw data by time (e.g., annual average), space (e.g., sum by region), or variables (e.g., multi-sensor data).
- **Fine-grained retention:** If users need to interact with details (e.g., "click to view hourly temperature fluctuations on a certain day"), they need to retain the original high-resolution data (e.g., sensor data recorded every minute) and optimize performance through technical means (e.g., dynamic loading).
- **Multi-granularity:** Commonly used in visualizations that support "zoom-drill-through" interactions (e.g., when a map is scaled from a country view to a city view, the data are automatically switched from a provincial aggregation to a municipal breakdown).

Practical method:

- Perform multi-granularity preprocessing of data in advance (e.g., generate a three-level aggregation table of "year–month–day" based on interactive design documents (e.g., "users need to support year/month/day three-level time filtering").
- Perform efficient granularity conversion using database aggregation functions (e.g., SQL's GROUP BY and AVG()) or programming tools (e.g., Python's pandas' resample()).

2. Standardized Coding of Space and Time

If the visualization includes spatiotemporal dimensions (such as dynamic temperature distribution on a map and time series map), ensure that the spatial coordinates (such as latitude and longitude) and timestamp meet the requirements of the visualization tool:

- **Spatial coding:** Unify geographic coordinates into standard projections (such as WGS84 latitude and longitude or Web Mercator projection), and convert spatial regions (such as provinces, and river basins) into boundary files (such as GeoJSON, Shapefile) that can be recognized by visual tools.
- **Time encoding:** Convert timestamps to formats supported by the visualizer (e.g., date object for JavaScript and datetime object for Python) and sort them in chronological order (to ensure that the data are presented in the correct chronological order during dynamic playback).
- **Spatial–temporal correlation:** If the data contain both spatial and temporal information (such as spatial–temporal monitoring data of sensor networks), a "spatial–temporal index" (e.g., through the composite key of latitude and longitude + timestamp) needs to be constructed to quickly retrieve specific times empty range data (Mahmood, *et al.*, 2019).

3. Variables: Mapping Predefinitions for Graphic Elements

According to the visualization goals in the requirements definition (e.g., "use colors to indicate temperature levels and lines to show trends"), define the mapping relationship between key variables and graph attributes (i.e., "data fields → graphical encoding") to ensure that subsequent visualization designs can be directly invoked:

- **Numerical variables:** These are usually mapped as color gradients (e.g., temperature → blue → red gradients), magnitude (e.g., concentration → dot size), altitude (e.g., elevation → histogram height), or position (e.g., coordinate axis values).

- **Categorical variables:** These are usually mapped by color categories (e.g., soil type → sand/clay/loam corresponding to different colors), shapes (e.g., different species with different markers), or grouping (e.g., separated by category).
- **Time variables:** These are typically mapped as x-axis (time series plot), animation frame (dynamic playback), or timeline filter controls.
- **Spatial variables:** These are typically mapped as geographic coordinates (points/lines/face locations on a map) or spatial grids (such as heat maps of raster data).

Practical method:

- Document this relationship through a "Variable-Code Mapping Table" document (e.g., "variable 'temperature' → color mapping (colormap: viridis), variable 'soil type' → categorical color (sand soil = #FF0000, clay soil = #00FF00)") for reference by designers and developers.
- In interactive design, leave room for flexibility in mapping adjustments (such as allowing users to customize color schemes or replace graphical properties).

6.3.4. *Validation Optimization of Data-Visual Mapping: Ensuring "Scientific Reliability"*

The goal of data adaptation is to ensure that the processed data accurately and efficiently drives visualizations and supports the reliability of scientific conclusions. Therefore, it is necessary to verify and optimize after the adaptation is completed to check the matching degree of the data with the visualization goal and potential problems.

1. **Data Integrity Verification:** Check whether the key variables are complete (e.g., whether temperature and humidity are included in the visualization target), whether the data subset covers the scope of user requirements (e.g., whether there are no omissions in 2020–2023), and whether missing values and outliers are reasonably handled (e.g., marking "incomplete data" or keeping outliers and explaining the reason) (Kosara, 2021).
2. **Data-Graph Mapping Verification:** Verify that the mapping of variables to graphical properties is as expected by rapid prototyping

(e.g., "Does the higher temperature really show the redder color?" and "Is the timeline in the correct order?") to avoid misunderstanding caused by mapping errors (such as color gradient reversal causing users to misjudge high and low values) (Lam *et al.*, 2020).

3. **Performance and Interaction Adaptation Verification:** Check whether the data volume is suitable for the performance limitations of the visualization tool (e.g., millions of data points may cause rendering lag on the web, which requires aggregation or sampling) and whether the interaction (such as scaling and filtering) can respond within a reasonable time (e.g., the graph should be updated within 1 second after clicking filter). If performance is insufficient, the data need to be further optimized (e.g., reducing accuracy and reducing redundant fields) or adjusting the visualization form (e.g., replacing the scatter plot with a heat map) (Hansen, *et al.*, 2014).
4. **Scientific Logic Verification:** Scientists confirm that the processed data still support the analysis of the original scientific question (e.g., "Does the filtered subset of data retain associations between key variables?"). "Does standardization affect the scientific meaning of variables?" to ensure that the visualization results do not draw incorrect conclusions due to improper data adaptation (Wimberly, 2023).

6.3.5. *Data Adaptation: The "Data Engine" of Visualization Projects*

Data adaptation is a key link in scientific visualization projects — it inherits the scientific goals and user needs in the requirements definition and starts the visualization form design, interactive implementation, and technology implementation. Through precise data screening, rigorous cleaning and preprocessing, structured transformation and verification optimization, data adaptation transforms the original data into a visual driving force of "usable, easy to use, and sufficient," ensuring that the final work can not only accurately convey scientific information but also support efficient interaction and in-depth exploration. In the following chapters, we will further explore the selection of visualization forms, the design of interaction logic, and the construction methods of immersive scenes based on the adapted data.

Chapter 7

Creative Practice Case Analysis

7.1. Introduction: When Theory Becomes Practice — A "Living Case Library" and a "Source of Inspiration" for Visualizing Innovation

The core value of scientific visualization lies not only in its "functionality" as a technical tool but also in its "creativity" as an innovative practice. The interdisciplinary art of complex data logic and intuitive visual language is deeply integrated, and every successful visualization practice is a vivid answer to the question of how to make science see, understand, and remember. If we say that the first six chapters of scientific visualization are the "underlying logic" (basic theory), "technical support" (digital media language), "data foundations" (acquisition and processing), "core approaches" (coding and interaction), "immersive experiences" (interaction and immersion techniques), and "system processes" (project management), in this chapter, we reveal the transformation code of scientific visualization from "theoretical drawings" to "practical works" through a series of real and vivid practical cases and show the multiple possibilities of visual innovation in different fields and different scenarios (Munzner, 2022; Kosara, 2021).

These examples are not only "concrete displays" of scientific visualization techniques but also "innovative experimental fields" of interdisciplinary collaboration. They include both "hard-core exploration" for the cutting edge of scientific research (such as microsimulations of quantum state superposition and dynamic tracking of gene folding dynamics) and "soft science" for public understanding (such as climate metaphors for the

Earth's breathing and adventure narratives for the cell kingdom). There are classical practices in traditional science (such as multidimensional representation of climate change and dynamic monitoring of marine ecosystems) and innovative attempts at emerging cross-cutting scenarios (such as immersive installations that integrate art and science and interactive experiences driven by digital media technologies). Each case is like a unique "innovation puzzle," which together forms a "panorama" of scientific visualization — from which we can see that how does the same scientific data differ from different design perspectives (e.g., rigorous scientific narratives vs. interesting life metaphors), technology choices (e.g., static graphics vs. dynamic interaction vs. immersive experience), and user positioning (e.g., professional scientists vs. the general public) have different vitality (Wimberly, 2023; Lam *et al.*, 2020).

More importantly, they are "living textbooks" for scientific visualization. By analyzing their design ideas, technical implementations, and user feedback, we can intuitively understand "Why choose this visual coding?," "How can I improve exploration efficiency through interaction design?," and "Where is the balance between art and science?" Wait for the key questions. For example, a successful popularization case might lower the public's awareness of abstract concepts through anthropomorphic characters such as "carbon elves," while a cutting-edge scientific tool might rely on high-precision 3D modeling and real-time rendering to reveal the dynamics of the micro-world. An engineering visualization project for engineers might focus on precise comparison of multivariate data, while an educational visualization for teenagers would focus on a mixture of fun and storytelling. These differentiated practical strategies are the scientific visualization of "local conditions," "vary from person to person," the concentrated expression of innovative wisdom.

The selected cases in this chapter cover the main application scenarios of scientific visualization (scientific research support, popular science education, engineering practice, and art integration) and typical fields (life science, Earth science, physical science, and environmental science); each case are analyzed from the four dimensions of "background needs," "visualization solution," "technical highlights," and "innovation value," not only to show "what was done" but also to dig deeper into "why was done" and "what effect was achieved." Through these examples, readers can not only see the "final product" of scientific visualization but also understand the "design philosophy" behind it, i.e., how to bridge the gap between science and cognition through the visual language. Let data no longer be a cold symbol but a "living medium" that carries scientific wisdom and inspires enthusiasm for exploration.

It can be said that the analysis of creative practice in this chapter is not only the "practical verification" of the previous theoretical methods but also the "inspiration" for the future innovation of scientific visualization. It tells us that the charm of scientific visualization lies precisely in its eternal "rigorous" and "creative," to find a balance and breakthrough between "technique" and "art," "profession" and "Volkswagen" — and this is the driving force to promote scientific visualization and continue to empower human cognitive advancement.

7.2. Popular Science Project Case: A Visual Narrative That Makes Science "Alive"

The core goal of the science popularization project is to transform complex scientific knowledge into visual content that the public can "understand, remember, and be interested in" and stimulate the public's curiosity and desire to explore science through artistic expression, story-based narrative, and immersive experience. This type of project is usually aimed at non-professional groups (such as teenagers and ordinary citizens), emphasizing the balance between "scientific" and "popular" — not only to accurately convey the essence of scientific concepts but also to lower the cognitive threshold through intuitive visual language (such as graphic metaphors and dynamic presentations). The following is through four typical cases (two new ones) to analyze how popular science visualization realizes the "temperature transfer of knowledge" through creative design.

Case 1: "Earth Breathing" — Global Climate Change Dynamic Visualization Science Popularization Exhibition

Project background: In response to the "dual carbon" goal, a science and technology museum planned the theme exhibition "Climate Change and the Future of Mankind," with the goal of visually showing the current situation, causes and impact of global climate change to the public (especially young people) through visual means, and conveying the scientific concept of "human activities affect the climate, and climate acts against humans."

Scientific Questions and Visualization Goals:

- **Scientific Questions:** These include the drivers behind rising global temperatures, frequent extreme weather (e.g., greenhouse gas emissions and deforestation), and their regional differences.
- **Visualization objectives:** (1) Demonstrate long-term trends in global temperature changes over the past 100 years, (2) reveal the association between greenhouse gas (e.g., CO_2) concentrations and air temperature, (3) show the differences in the impact of climate change (e.g., melting glaciers and sea level rise) in different regions (e.g., polar regions, coastal cities), and (4) stimulate public thinking about low-carbon life.

Data adaptation and visualization design:

- **Data selection:** Data from authoritative scientific institutions (such as NASA global temperature data, NOAA atmospheric CO_2 concentration monitoring data, and IPCC sea level rise prediction data) are selected, covering the time range (1900 to present), spatial scope (global land area and ocean), and key variables (air temperature, CO_2 concentration, glacier area, and sea level height).
- **Core visualization forms:**
    - **Dynamic timeline projection:** Using the Earth panoramic model as the background, the global average annual temperature change curve is displayed through a dynamic timeline (1900–2023) — the temperature data are mapped to different regions of the Earth's surface with color gradients (blue → red), and the audience can intuitively see that the polar regions (such as the Arctic Circle) are warming much higher than the Equator (such as the Sahara region of Africa), echoing the scientific conclusion of the "polar amplification effect." As the timeline slides, the Earth model simultaneously displays the glacier coverage for the corresponding year (simulating glacier retreat through three-dimensional topographic height changes), such as the dynamic process of the Greenland ice sheet thickness reduction between 1980 and 2020.
    - **CO_2–temperature correlation bubble plot:** The annual global atmospheric CO_2 concentration (X-axis) and annual average temperature (Y-axis) are plotted into a scatter plot, with each point representing a year, and the size of the dots maps the number of global extreme heat events (such as days exceeding 40°C) in that year. Through the animation of the data point generation process of nearly 50 years, the audience can clearly observe the positive correlation

trend between the increase in CO_2 concentration (increase in X-axis value) and the increase in temperature (increase in Y-axis value), and at the same time, the increase in the size of points suggests that the frequency of extreme weather is increasing simultaneously, intuitively conveying the scientific logic of "greenhouse gas emissions exacerbating climate anomalies."

- **Immersive scene experience:** This sets up the "Future City" VR exhibition area — after wearing the VR equipment, visitors can choose different emission reduction scenarios (such as "current emissions," "moderate emission reduction," and "deep emission reduction") to observe the impact of global sea level rise on coastal cities (such as Shanghai and New York) in 2100: Under the deep emission reduction scenario, the urban coastline is basically stable. Under the current discharge scenario, the water level of the Huangpu River submerges the Bund complex, and with sound effects (such as waves and alarms) and text prompts (such as "If you don't act, your hometown may be like this"), the public's perception of the urgency of emission reduction is strengthened.

Interaction Design Highlights:

- Visitors can independently select areas of interest (e.g., "Antarctica" and "Amazon rainforest") through touch screens to view temperature change curves and biodiversity impact data (e.g., species extinction rates) in the region.
- Set up the "Carbon Footprint Calculator" mini-game — the audience enters their daily behavior (e.g., "drive 5 times a week" and "use the air conditioner for 20 hours"). The system calculates an individual's annual carbon emissions in real time (e.g., "equivalent to cutting down 3 trees") and recommends corresponding low-carbon behaviors (e.g., "changing to a bus" or "raising the air conditioning temperature by 1°C"), transforming abstract carbon emission concepts into perceptible personal actions.

Project results: The exhibition received more than 100,000 visitors within half a year of opening, and the on-site questionnaire showed that 85% of the young audience could accurately describe the "relationship between greenhouse gases and temperature," and 70% of parents reported that their children took the initiative to discuss "how to reduce family carbon emissions" after returning home, achieving the dual goal of "popularizing scientific knowledge + guiding behavioral awareness."

Case 2: "Cell Kingdom Adventure" — Microbiology Visualization Science APP

Project background: In response to the teaching difficulties of "cell structure and function" in junior high school biology courses (such as students finding it difficult to imagine the dynamic process of the microscopic world), an educational technology company has developed a popular science app for teenagers, with the goal of helping students "enter" the inside of cells and understand the division of labor and cooperation of organelles and the microscopic mechanism of life activities through gamified interaction and three-dimensional visualization.

Scientific Questions and Visualization Goals:

- **Scientific questions:** These include the structure, function, and synergy of various organelles (e.g., mitochondria, chloroplasts, and Golgi apparatus) in cells (e.g., energy conversion and protein synthesis).
- **Visualization objectives:** (1) Show the three-dimensional overall structure of cells and the spatial distribution of each organelle, (2) dynamically demonstrate the functional processes of organelles (such as how mitochondria perform aerobic respiration to produce ATP and how ribosomes synthesize proteins), and (3) stimulate students' interest in microbiology through interactive exploration.

Data adaptation and visualization design:

- **Data sources:** These are based on real cell biology research results (e.g., high-definition images of organelles under electron microscopy and molecular biology experimental data), combined with educational goals to simplify complex mechanisms (e.g., simplifying the ATP synthesis process into a three-step process of "electron transport chain → proton pump → ATPase power generation") to ensure scientific accuracy while lowering the cognitive threshold.
- **Core visualization forms:**
  - **Three-dimensional immersive cell roaming:** After entering the APP, users first see an enlarged "cell kingdom" — the cell membrane serves as a transparent boundary, and colored organelles are suspended inside (mitochondria are like "power plants" in the shape of rods, chloroplasts are like "solar panels" in the form of flat sacs, and Golgi apparatus are like "courier stations" in the form of stacked vesicles). Users can rotate the viewing angle by swiping their fingers (360° view of the whole cell) and zoom with two fingers to focus on

specific organelles (e.g., zooming in on mitochondria to observe the "crest" structure formed by the folding of their inner membranes).

- **Functional animation demonstration:** Click on any organelle to trigger a dynamic demonstration — e.g., after clicking on a mitochondria, the screen switches to its interior, showing the transfer of electrons through the protein complex (simplified to flashing globules) on the inner membrane, driving protons (H^+) from the matrix pump to the membrane gap (represented by blue particles) to form an electrochemical gradient. The protons then flow back to the matrix (green channel) through the ATPase enzyme, promoting ADP and Pi to synthesize ATP (golden globules are generated and fly out), with the voice explanation ("This is the cell's 'energy factory,' producing hundreds of millions of ATP per second!"). Similarly, click ribosomes demonstrate how mRNA carries genetic information (green chains) and guides amino acids (colored globules) to assemble into proteins (long chain structures).
- **Interactive task challenge:** Design the "Cell Factory Operation" mini-game — the user plays the role of a "cell administrator" and needs to adjust organelle work according to cellular needs (e.g., "muscle cells need more energy"): Drag glucose molecules to mitochondria (trigger more ATP production) and transport amino acids to ribosomes (synthesize specific proteins). In the event of an error (such as throwing waste into the Golgi apparatus), the cells will send out an "alarm" (such as "languishing" due to lack of energy), reinforcing students' understanding of organelle function through gamified feedback.

Interaction Design Highlights:

- **Multimodal feedback:** In addition to visual animations, each organelle demonstration is accompanied by interesting audio commentary (such as mitochondrial self-introduction: "I am the 'power bank' of cells, without me, your muscles are hard to jump!"), key steps accompanied by sound effects (such as the "ding-dong" sound when ATP is generated).
- **Layered learning:** Basic mode only shows the appearance of organelles and simple functions, and advanced mode (for students interested in biology) unlocks more details mechanisms (e.g., specific protein names of electron transport chains and codon pairing processes for protein synthesis);

- **Social Sharing:** After completing interactive tasks, users can generate screenshots of "My Cell Kingdom" (showing customized organelle layouts and functional status) and share them on social media to stimulate learning competition among peers.

Project effectiveness: After the APP was promoted in pilot schools, students generally reported that "cell structure is no longer abstract" and "the working scene of organelles can be imagined when doing questions," and teachers commented that it "transformed the static illustrations in the textbook into a dynamic and operable microscopic world, which significantly improved the teaching efficiency."

Case 3: Space Seed Adventures — Immersive Science Popularization Exhibition of Aerospace Breeding

Project background: Space breeding (carrying seeds into space through returnable satellites, or spacecraft, using microgravity, strong radiation and other special environments to induce genetic mutations, and cultivating new crop varieties with high yield and disease resistance) is the cutting-edge technology of modern agriculture, but for the public, questions such as "How does the space environment change seeds?" and "Are mutated seeds safe?" often cause misunderstandings. An Academy of Agricultural Sciences cooperated with media institutions to create an immersive popular science exhibition "Space Seed Adventures," which unveils the scientific veil of space breeding through simulated space environments, interactive experiments, and storytelling narratives.

Scientific objectives: These explain the basic principles of aerospace breeding (the effects of microgravity and cosmic radiation on seed genes), demonstrate the actual results of aerospace breeding (such as high-yield rice and insect-resistant wheat), and eliminate the public's confusing perception of "genetically modified" (aerospace breeding is natural mutation screening, not artificial gene editing).

Visual Design:

- **Space environment simulation:** The "space capsule experience area" is set up at the entrance of the exhibition, allowing the audience to "experience" the journey of seeds in space through VR equipment — the screen shows the shocking scene of satellite launches and then switches to a microgravity environment (the audience "floats" in the cabin to observe the suspended state of the seeds in a weightless state), with strong radiation special effects (the screen flashes blue light, simulating the process of cosmic rays penetrating the seeds).

- **Seed variation visualization:** This compares electron microscope images (1000 × magnification of seed cell structure) with 3D animation to show the difference in thickness between ordinary seeds and aerospace seeds (aerospace seed cell walls are thicker and more stress-resistant) and changes in chloroplast distribution (aerospace seeds have denser chloroplasts and higher photosynthesis efficiency) through electron microscopy images (1000 times magnified seed cell structure) and 3D animations, and dynamic growth simulation (ordinary seeds have slow germination and are prone to leaning over, while space-traveled seeds have developed root systems and sturdy stems) to visually present the advantages after mutation.
- **Interactive experimental bench:** Audiences can operate simulated experiments (such as using the "radiation intensity adjustment knob" to observe the effect of different radiation doses on seed germination rate, or selecting high-yielding individuals from mutant seeds through the "genetic screening game"), and the experimental results generate data charts in real time (e.g., "When radiation dose X, the germination rate increases by 20%"), which is interpreted by popular science explainers based on scientific principles (e.g., "Low dose radiation may activate silencing genes and induce beneficial mutations").
- **Achievement display area:** This displays the physical and comparative data of real aerospace breeding crops (such as space pepper and space corn) (such as "ordinary corn yields 800 pounds per mu and aerospace corn yields 1200 pounds per mu") and interviews with growers through short videos ("after planting aerospace corn, the amount of pesticides reduced by 30%") to strengthen the actual value of science and technology.

Case 4: Carbon Cable Future — Children's Science Picture Books and AR Extension of Carbon Cycle

Project background: The carbon cycle (such as the absorption and release of carbon dioxide in the atmosphere, oceans, and terrestrial ecosystems) is fundamental to understanding climate change, but for children, "how invisible carbon affects the planet" is too abstract. An environmental protection organization has teamed up with children's education experts to develop a series of popular science picture books,

"Carbon Cable Future," and supporting AR applications, allowing children to "see" the journey of carbon through childlike stories and augmented reality technology.

Scientific goal: It is to convey to children aged 7–12 the three core concepts of "natural carbon cycle processes (such as plant photosynthesis absorbing CO_2 and animal and plant respiration releasing CO_2)," "how human activities (such as burning fossil fuels) break the balance," and "daily actions to reduce carbon emissions (such as planting trees and saving electricity)."

Visual Design:

- **Picture book storytelling:** This features the anthropomorphic "carbon elf" as the protagonist, telling the story of its atmospheric nature, the story of encountering different partners when traveling (such as the big tree "huhu" absorbing it through photosynthesis, and the car "beeping" emitting a lot of carbon and making it "crowded and uncomfortable"). Each page uses exaggerated illustrations (such as transparent carbon bubbles floating in the atmosphere, deforested forests exposing bare land) to show the key links of the carbon cycle, and the text descriptions rhyme with simple rhymes (such as "Big trees stretch their waists, carbon elves jump; Cars emit black smoke, carbon elves escape").
- **AR interactive extension:** Scan specific pages in the picture book (such as "forest scene" and "power plant chimney") with your mobile phone to trigger 3D animation — when scanning the forest page, the virtual tree root system extends underground, showing the process of rhizobia carbon sequestration, and children can click on the leaves to "collect" the floating carbon bubbles. When scanning the power plant page, the animation demonstrates the chemical reaction of CO_2 produced by coal combustion (molecular model disassembly and recombination) and prompts that "if we turn on the air conditioner for one hour less, we can reduce XX grams of carbon emissions."
- **Offline experience activities:** Supporting the "Little Carbon Detective" parent–child workshop, children use pH test strips to measure the acidity of rainwater (acid rain caused by carbon emissions) and use simple devices to measure the carbon sequestration capacity of different plants (e.g., comparing the carbon dioxide absorption capacity of the Philodendron plant with that of the Cactus), and transform picture book knowledge into hands-on practice to deepen understanding.

Common experience and enlightenment of popular science project cases from "Breathing the Earth" to "Carbon Cable Future" and successful science visualization projects often have the following core characteristics:

- **Metaphor-driven cognitive transformation:** This concretizes abstract scientific concepts through life-like metaphors (such as earth breathing, carbon elves) to lower the public's cognitive threshold (Zhang *et al.*, 2023).
- **Multi-sensory collaborative experience:** This integrates multi-channel information such as visual (dynamic light and shadow/3D animation), auditory (ambient sound effects/story dubbing), and tactile (interactive operation/physical exhibits) to strengthen memory and emotional resonance (Hansen, *et al.*, 2014).
- **Balance between "scientific accuracy" and "popular and fun":** This strictly follows scientific facts (such as non-GMO in aerospace breeding and chemical principles of carbon cycle) but avoids boring preaching through storytelling (role-playing, childlike illustrations) (Sacha, *et al.*, 2020).
- **Participation and action guidance:** This promotes the transformation of the public from "knowing" to "acting" through interactive experiments (simulated radiation experiments), offline activities (carbon detective workshops), or daily recommendations (action checklists to reduce carbon emissions) (Kosara, 2021).

These cases prove that science visualization is not only a "translator" of scientific knowledge but also a "catalyst" to stimulate public interest in science and cultivate rational thinking. In the future, with the development of artificial intelligence (such as personalized popular science content generation), virtual reality (such as immersive scientific exploration), and Internet of Things (such as real-time monitoring of household carbon footprints) technology, popular science visualization will further evolve in the direction of "personalization, immersion, and life" so that science can truly become a "good friend" in everyone's life (Andrienko *et al.*, 2022).

The core goal of the science popularization project is to transform complex scientific knowledge into visualization content that is easy for the public to understand, stimulate public interest in science, and convey scientific ideas and methods through intuitive graphics, dynamic interaction, and immersive experience. This type of project is usually aimed at non-professional groups (such as teenagers and ordinary citizens), so it is necessary to focus on the popularization of information, visual appeal, and the interest of interaction on the basis of scientific rigor. The following analyzes how scientific visualization in popular science projects achieves the dual goals of "knowledge dissemination" and "public participation" through creative design.

7.3. Scientific Support Cases: A Visual Engine from Data Exploration to Scientific Findings

The core goal of scientific support visualization projects is to provide researchers with exploratory data. The essence of "cognitive tools" to validate hypotheses and discover patterns is to transform scientific data into interactive and analytical visual expressions, helping researchers to break through the limitations of traditional analytical methods and gain insight into the scientific mechanisms hidden in complex data. This type of project is usually aimed at specialized researchers (e.g., physicists, biologists, and geoscientists) and emphasizes "professionalism," "interactivity," and "deep analytical capabilities" — the need to accurately adapt scientific data from a particular field (e.g., molecular structure, astronomical observations, and geological exploration data, in turn, need to be supported by dynamic visualization and interactive technologies to support multilevel exploration from macro trends to microscopic details. Here are four typical examples (two new ones) of how scientific visualization acts as an "engine" to drive scientific discovery.

Case 1: Interactive Visualization Tool for Quantum State Superposition
Project Background: Superposition states in quantum mechanics, such as the superposition of electrons in multiple locations at the same time, are one of the core properties of the microscopic world, but their abstraction — which cannot be easily described in classical physics' position–velocity — makes them difficult for students and research novices to understand. A quantum physics lab has developed an interactive visualization tool called "quantum state superposition," which interacts with users through dynamic simulations to help users visualize the probability distribution of quantum states and measure the collapse process.

Scientific objective: To demonstrate the principle of superposition in quantum states (such as superposition in which an electron exists in multiple locations at the same time). Measurements lead to collapse (where electrons appear randomly at a certain location after measurement) and entanglement correlations between multiple qubits (e.g., the state of two entangled qubits changes in tandem).

Visual Design:

- **Core metaphor:** This maps the probability distribution of quantum states as electron clouds (representing the probability density of electrons appearing in a region with translucent color blocks and the color depth corresponds to the probability of high or low), and the superposition state shows the overlap of multiple electron clouds (such as when electrons appear at both point A and point B and the superposition of electron clouds in A and B regions is enhanced). The measurement collapse is triggered by a "click on the measurement" button — the electric cloud instantly shrinks to a specific location (such as point A), and the probability of the simulation measurement is determined.
- **Interaction:** Users can adjust the parameters of the superposition state (such as the coefficient of superposition in two locations) through a slider to observe the shape change of the electronic cloud in real time (the greater the coefficient, the more dense the electronic cloud at the corresponding location). The relationship between abstract mathematics and object diagram is established through multi-view linkage (the left side shows mathematical formula and probability, and the right side shows electron cloud dynamics). This supports the addition of multiple qubits to observe the synchronous movement of two electron clouds in entangled states (e.g., as one electron cloud contracts and the other collapses to an associated location).
- **Instructional assistance:** Built-in "Frequently Asked Questions" pop-ups (e.g., "Why do electrons appear in a given location after measurement?" and "What is the difference between superposition and classical probability?"), combined with concise scientific explanations, assists the user in the transition from intuitive experience to theoretical understanding.

Case 2: Dynamic Simulation Platform for Global Ocean Circulation
Background: Ocean circulation (e.g., the Gulf of Mexico Current and the Antarctic Circumpolar Current) has a profound impact on global climate (e.g., heat transport and carbon cycling), but its complex dynamics (e.g., wind-stress driven and salt-temperature coupling) are difficult to visualize

with single numerical simulations. An oceanographic institute has developed a dynamic simulation platform, Global Ocean Circulation, which uses high-resolution numerical analog data and three-dimensional visualization technology to help scientists analyze the spatial structure, temporal variation, and climate correlation.

Scientific objectives: These are to demonstrate the flow paths of major global ocean circulations (such as the distribution of surface warm and deep cold currents), the coupling relationship between flow velocity and temperature/salinity (e.g., a warmer current carrying high warm water and a colder current sinking to drive deeper circulation), and the potential impact of human activities (e.g., warming of seawater due to greenhouse gas emissions) on circulation patterns.

Visual Design:

- **Multivariate integration:** Numerical simulation of the output of grid data (including flow speed, temperature, salinity, and depth) is displayed by color (temperature: blue to red indicates low to high temperature), flow lines (direction and density indicate flow speed and direction), and equivalence surfaces (vertical areas corresponding to salinity thresholds). For example, the warm core of the Gulf of Mexico Current is represented by red high-speed streamlines, and the associated deep cold current is represented by blue low-speed streamlines and dark is salinity surfaces (high salinity cold water).
- **Dynamic time series simulation:** This supports sliding timelines (almost 50 years of data backtracking) to observe long-term changes in circulation (such as the slowing of the Arctic circulation due to melting sea ice). Animation shows seasonal fluctuations (such as the process of reversing the flow of the Indian Ocean circulation when the monsoon strengthens in the summer).
- **Interactive analysis tools:** Scientists can frame a specific region (such as the North Atlantic warmer current and the European climate associated region) and view the flow rate-temperature dispersion maps of that region, vertical cross-section diagrams (stratification of velocity and salinity at different depths), or analysis of mass transport paths (e.g., efficiency of transport of nutrients from the equator to higher latitudes) by particle tracking techniques (release of virtual particles following circulation).

Case 3: Massive Parallel Visualization Platform for Galaxy Evolution

Project Background: The formation and evolution of galaxies (such as the birth and death of stars and mergers among galaxies) is one of the core problems of cosmological research. However, numerical simulations, such as models of galaxies containing hundreds of millions of stars, generate extremely large amounts of data (up to terabytes in a single simulation) and involve multiple physical process couplings (such as gravitational interactions, gas cooling, and star formation). A research center for astrophysics has developed a parallel visualization platform, "Galactic Evolution," which supports researchers' multilevel exploration from the large-scale structure of the universe to the details of the inner galaxy through distributed computing and interactive analysis.

Scientific objectives: These include demonstrating the morphological evolution of galaxies (e.g., how spiral galaxies transition to elliptical galaxies), space-time distributions of star formation (e.g., stellar birth rates in galactic centers and spheres), and dynamics during galactic mergers (e.g., tidal tails caused by gravitational perturbations, such as explosive star formation caused by gas compression).

Visual Design:

- **Multi-resolution rendering:** This is for massive amounts of galaxy data (including different components of stars, gas, and dark matter). Using adaptive grid refinement technology, local areas of interest to the user, such as the collision interface of galaxy mergers, are rendered in high resolution (showing the trajectory and gas density of individual stars). Large-scale structures of the background (such as dark matter filaments in the cosmic web) are simplified in low resolution, balancing computational efficiency with detail.
- **Multiphysics superposition:** This represents star formation rates (density of light points in blue). The gas temperature (red to blue indicates high to low temperature), and the dark matter distribution (translucent grey grid) are superimposed through different visual channels, revealing multi-process coupling mechanisms (e.g., the density of the gas increases after cooling, triggering star formation, while dark matter gravity dominates the overall motion of the galaxy).
- **Interactive scene switching:** Scientists can use a timeline slider to retrace the evolution of galaxies (from the collapse of gas clouds in the

early universe to the maturation of modern galaxies) or switch observational perspectives (from global cosmic scale focus to the formation of individual stars). This supports the addition of tags (such as marking the location of the star's exploding supernova) and the generation of dynamic trajectories (showing the path of the star throughout the galaxy and its final destination).

Common values and technical implications of scientific support cases from quantum state superposition to galaxy evolution and successful scientific visualization tools typically have the following core features:

- **Professional suitability:** An in-depth understanding of the properties of domain-specific scientific data (e.g., probability amplitudes of quantum states, amino acid sequences of proteins, and multi-component simulations of galaxies) to ensure that visual coding (e.g., colors, streamlines, and three-dimensional models) accurately reflects the nature of science;
- **Interaction-driven exploration:** Through dynamic simulation, parameter tuning, and multi-view interactions (such as showing both macro galactic morphology and micro star formation rates), researchers are supported in closed-loop studies from hypothesis formulation → data validation → new problem discovery.
- **High-performance computing support:** This is for large-scale data (such as global grids of ocean circulation, hundreds of millions of particles modelled on galaxies), and rely on parallel computing (GPU acceleration, distributed rendering) and data compression techniques (such as rendering only high-resolution details of key areas) to balance visualizations with real-time performance;
- **Cross-scale integration:** This links microscopic mechanisms (such as the hydrogen bonding of proteins and the distribution of electron clouds in quantum states) to macroscopic phenomena (such as climate effects of ocean circulation, cosmological significance of galaxy mergers) to help scientists build systematic understanding.

These cases demonstrate that scientific visualization is not only a "show window" for data but also an "acceleration engine" for scientific discovery by transforming complex data into interactive, analyzable visual language that allows researchers to "see" patterns that are difficult to capture with traditional methods, thus facilitating the leap from theoretical hypotheses to empirical findings. In the future, with the development of artificial intelligence (such as automatic identification of critical intermediate states in folding paths), virtual reality (such as immersive exploration of the inner structure of galaxies), and quantum computing (processing of very large analog data), research visualization will further break through cognitive boundaries and become an indispensable "intelligent partner" in scientific research.

Case 4: "Wind And Cloud Tracking" — An Interactive Visualization Platform For Typhoon Track And Intensity Prediction

Project background: Typhoons (hurricanes) are one of the most serious natural disasters affecting coastal areas around the world, and accurate prediction of their path and intensity is crucial for disaster prevention and mitigation. Traditional meteorological forecasting relies on massive data output from numerical models (such as global atmospheric circulation field, temperature field, and humidity field every 6 hours), but researchers and forecasters face challenges in analyzing these data: The spatiotemporal dynamic correlation of multiple variables (such as wind speed, air pressure, sea temperature) is complex, and it is difficult to intuitively discover "which combination of factors is most likely to cause sudden changes in typhoon intensity" and "how environmental conditions in specific sea areas affect path shifts." A national meteorological research institute and a visualization team have developed a "wind and cloud tracking" platform, aiming to assist in the optimization of typhoon prediction models and the mining of scientific mechanisms through interactive visualization.

Scientific Questions and Visualization Goals:

- **Scientific question:** What are the key environmental factors (e.g., sea surface temperature, vertical wind shear, and water vapor content) that affect the dynamics of typhoon intensity (e.g., maximum wind speed) and track (e.g., direction/velocity)? Is there a differentiated mechanism for typhoon formation and evolution in different sea areas (e.g., Northwest Pacific vs. North Atlantic)?
- **Visualization objectives:** (1) Display multivariate spatial-temporal data (such as wind speed, atmospheric pressure, sea temperature, and vertical wind shear) of the entire life cycle of the typhoon (from generation to dissipation), (2) reveal the correlation pattern between environmental factors and typhoon intensity/track (e.g., "typhoon intensity

is likely to increase when the sea temperature is > 28°C and the vertical wind shear < 10m/s"), and (3) support researchers to interactively explore the details of specific typhoon cases, and assist in the parameter adjustment and mechanism hypothesis verification of prediction models.

Data adaptation and visualization design:

- **Data selection:** This integrates multi-source meteorological observations and numerical simulation data, including the following:
  - Typhoon optimal path dataset (providing time-series information such as typhoon center position, maximum wind speed, and minimum air pressure, with a temporal resolution of 6 hours and a spatial resolution of 0.5° × 0.5°);
  - Re-analysis of meteorological data (such as global sea surface temperature, 850/200 hPa wind field, vertical wind shear, relative humidity, temporal resolution of 1-hour, and spatial resolution of 0.25° × 0.25°) provided by ERA5.
  - **Coverage time range:** This includes all named typhoons (about 500 cases) in the Northwest Pacific typhoon season (June–October) in the past 20 years (2003–2022).

- **Core visualization forms:**
  - **Three-dimensional spatiotemporal cube navigation:** This treats each typhoon case as a "cube," with the X/Y axis as the spatial dimension (latitude and longitude grid), the Z-axis as the time dimension (the time series from typhoon generation to dissipation), and each voxel (small cube element) in the cube encodes multivariable values (such as wind speed, sea temperature, and vertical wind shear). Researchers can quickly locate specific typhoons (such as Super Typhoon Haiyan in 2013) through rotating cubes and focus on key periods or areas through slicing operations (such as viewing the sea temperature distribution of typhoons at that time with a fixed time section or observing wind speed changes in the path with a fixed spatial area).
  - **Multivariate correlation heat map matrix:** For each typhoon case, extract the environmental factor data (such as sea temperature, vertical wind shear, and water vapor content) and typhoon intensity indicators (maximum wind speed and air pressure) every 6 hours on its path, and generate a correlation coefficient matrix heat map between

variables (color shades indicate the correlation intensity, such as red for strong positive correlation and blue for strong negative correlation). Researchers can click on any variable pair in the matrix (such as "sea temperature — maximum wind speed") to automatically generate the scatter plot + regression line of the variable pair and intuitively observe the quantitative relationship of "how many meters per second the typhoon wind increases on average for every 1°C increase in sea temperature."

- **Dynamic Path-Environment Overlay:** This overlays the historical path of typhoons on the map (represented by colored lines, and color shades map wind speed intensity) and dynamically displays the environmental factor values of each location on the path (e.g., bubble size represents sea temperature and bubble color represents vertical wind shear). Users can click on any point on the path to pop up the detailed data panel of that moment (e.g., "Time: 2022-09-01 12:00, Location: 15°N 125°E, Maximum Wind Speed: 65 m/s, Sea Temperature: 30.2°C, Vertical Wind Shear: 8 m/s") and compare the environmental field of the surrounding area during the same period (such as the distribution of warm and humid airflow on the east side of the typhoon center by zooming in on a small area).

Interaction Design Highlights:

- **Table example screening and comparison:** Typhoon cases can be filtered by intensity level (e.g., tropical storm/strong typhoon), generated sea area (e.g., South China Sea/ocean east of the Philippines), season (e.g., August vs. September), and multi-view synchronous comparison (e.g., side-by-side display of the paths and environmental fields of two similar typhoons) to assist in discovering common laws.
- **Parametric hypothesis verification:** Researchers can manually adjust the threshold of environmental factors (e.g., "only a typhoon case with an sea temperature > of 29°C and vertical wind shear < 12 m/s"), and the platform can screen eligible typhoon subsets in real time and update the visualization results (such as generating a statistical curve of intensity change for this subset) to quickly verify the hypothesis that "specific environmental conditions necessarily cause strong typhoons."
- **Model output integration:** By connecting with the real-time output data of typhoon numerical forecast models (such as the predicted wind speed field of ECMWF and GFS models), researchers can superimpose the prediction path and environmental field on historical cases, and

compare the prediction bias of the model (e.g., "the model underestimated the vertical wind shear in a certain sea area, resulting in an error in the prediction of path offset") to assist in model optimization.
- **Project results:** The platform has been applied to the daily business and scientific research projects of the Typhoon Forecasting Center of the China Meteorological Administration, helping researchers publish three SCI papers within two years (such as "Research on the Threshold of Environmental Factors for Sudden Changes in Typhoon Intensity in the Northwest Pacific"), and directly supported the forecaster's path correction prediction of the super typhoon "Dusuri" in 2023 (through the comparison of environmental fields of similar historical cases, and pay attention to the abnormal areas of vertical wind shear in advance).

Case 5: "Protein Universe" — A Multi-Scale Visual Analysis Tool for Protein Interaction Networks

Project Background: In the field of life sciences, proteins are the core executors of cellular functions, and the interactions between proteins (such as enzyme-substrate catalysis, receptor-ligand binding) constitute a complex molecular network, and their dynamic changes are closely related to disease occurrence (such as cancer and Alzheimer's disease) and drug target discovery. Traditional studies use high-throughput experiments (e.g., yeast double hybridization, mass spectrometry to obtain protein interaction data), but these data are usually large (containing tens of thousands of protein nodes and hundreds of thousands of edges), complex structures (modular clusters, key hub nodes), and lack an intuitive dynamic correlation display. A biomedical research institute has collaborated with a visualization laboratory to develop the "Protein Universe" tool, which aims to help researchers explore the topology, functional modules, and disease associations of protein networks through multi-scale visualization.

Scientific Questions and Visualization Goals:
- **Scientific question:** Is there a specific topological pattern in a protein interaction network (such as tightly connected "functional clusters" or "pivot proteins" that connect multiple clusters)? How do these patterns correspond to functional pathways (e.g., metabolic pathways, signaling) within cells? What are the location and connectivity characteristics of disease-related proteins (such as cancer markers) in the network?

- **Visualization objectives:** (1) Demonstrate the global topology of large-scale protein interaction networks (e.g., node degree distribution and clustering coefficient), (2) identify and visualize functional modules (e.g., protein clusters associated with metabolic pathways) and key nodes (e.g., pivot proteins) in the network, (3) support researchers to interactively explore the neighborhood relationships, dynamic changes (e.g., network differences in different tissues/disease states), and associations with disease phenotypes of specific proteins.

Data adaptation and visualization design:

- **Data selection:** Integrate multi-omics experimental data with public databases (e.g., BioGRID and STRING), including the following:
  - Protein interaction data (about 100,000 edges, connecting about 8,000 protein nodes, each edge is marked with the type of interaction (such as physical binding and co-expression), confidence score (0-1));
  - Protein function annotation data (e.g., GO term classification: metabolism, signal transduction, and DNA repair);
  - Disease-related data (e.g., cancer-related differentially expressed proteins in the TCGA database);
  - **Coverage:** The core interaction network of the human proteome (focusing on key pathways associated with disease).

- **Core visualization forms:**
  - **Multi-level force-oriented layout:** Using an improved force-oriented algorithm (considering node degree, edge confidence, and functional annotation similarity), the protein network is mapped into a two-dimensional plan — nodes represent proteins (size coding centrality, color-coded functional categories), and edges represent each other use (line thickness to encode confidence, transparency to encode interaction type). In the global view, users can observe the macroscopic clustering structure of the network (e.g., several closely connected "clumps" corresponding to metabolic pathway clusters). Local magnification allows for a clear view of the neighborhood relationships of individual proteins (e.g., a pivot protein connecting nodes with 10 different functional clusters).
  - **Dynamic modular clustering:** Algorithms (such as Louvain method) automatically identify functional modules in the network (such as "mitochondrial electron transport chain clusters" and "cell

cycle regulatory clusters") and mark the module boundaries with translucent areas of different colors. Users can click on the module to view the list of proteins it contains (categorized by functional annotation), key nodes within the module (e.g., the highest degree of "core protein"), and the connection relationship between the module and other modules (e.g., "metabolic cluster" and "signal transduction cluster" coupled through a pivotal protein).

 - **Disease-associated overlays:** Disease-related proteins (such as differential proteins upregulated/downregulated in cancer) are highlighted with special symbols (such as red star markers) or highlighted borders, and their topological features in the network are superimposed (e.g., "This cancer marker protein is located at the junction of multiple modules, connecting metabolic and apoptotic pathways"). Users can observe changes in the network during disease progression through the timeline slider (if there is dynamic data) (e.g., "A protein is connected to 5 modules in the early stage of cancer and only 2 modules in the advanced stage, indicating functional deterioration").

Interaction Design Highlights:

- **Multi-scale navigation:** This supports "global-local-single-node" three-level interaction — in the global view, the module area of interest can be zoomed and positioned by the mouse wheel, the layout can be adjusted by dragging the node in the local view (optimizing the clarity of local connections), and the detailed information panel will pop up after clicking on a single node (e.g., "Protein Name: TP53, Function: Tumor Suppressor, Degree Centrality: 85, Associated Diseases: Lung Cancer/Breast Cancer/Colorectal Cancer").
- **Dynamic filtering and path tracing:** Users can annotate by function (e.g., "Show only proteins related to signal transduction"), disease associations (e.g., "Show only cancer-related proteins"), and degree of node (e.g., "Show only degrees >). 20 pivot proteins") screening network to quickly focus on a subset of targets. After clicking on a protein, it automatically highlights the shortest path to other key nodes (e.g., "signaling pathway from epidermal growth factor receptor EGFR to downstream kinase AKT") to assist in the study of the regulatory mechanism of signaling pathways.
- **Cross-data correlation:** This integrates gene expression data (e.g., gene abundance value of RNA-seq) to map the color shade of protein

nodes to the expression level of their corresponding genes (e.g., red = high expression, blue = low expression), helping researchers correlate network topology with molecular expression dynamics (e.g., "a pivotal protein is highly expressed in tumor tissue and connects multiple resistance-related modules").

Project results: The tool has been adopted by many laboratories at home and abroad, playing a key role in cancer mechanism research and drug target screening — e.g., a team discovered a new "metabolic-immune crossover cluster" through the "protein universe," which contains a pivotal protein that has been confirmed to be a potential target for liver cancer immunotherapy, and the relevant results were published in *Cell Metabolism*. In addition, the tool helps researchers prioritize "vulnerable nodes" in the network (such as a small number of proteins connecting multiple functional clusters) in drug screening experiments, increasing target validation efficiency by 40%.

Common experiences of scientific research supporting visualization projects from the above five cases and successful scientific research support visualization usually have the following core characteristics:

- **Deeply adapt to scientific research scenarios:** These closely focus on specific scientific research problems (such as typhoon prediction and protein network analysis), select suitable data types (multivariate time series data and complex network data), and analysis goals (association discovery, mechanism verification, and hypothesis testing) to ensure that the visualization tool "solves real problems" (Munzner, 2022).
- **Deep integration of technology and science:** This is the combination of professional algorithms (such as typhoon numerical simulation and protein interaction network clustering), high-precision data (such as reanalysis of meteorological data and high-throughput experimental data), and flexible interaction design (such as multivariate screening and dynamic path tracing) not only ensures scientific rigor but also improves analysis efficiency (Ware, 2020).
- **Closed-loop support from exploration to verification:** This not only provides data display functions but also through interactive operations (such as adjusting parameters, comparing cases), quantitative analysis (such as correlation coefficient calculation, module feature statistics) and result export (such as screened data subsets and visual charts) to support researchers in the complete scientific research process from

"observing phenomena" to "verifying hypotheses" to "forming conclusions" (Heer & Shneiderman, 2021).

- **Scalability and Collaboration:** These support integration with scientific research toolchains (such as docking numerical forecast models and bioinformatics databases) and allow team members to work collaboratively (such as sharing filter conditions and labeling key findings) to promote interdisciplinary collaboration and knowledge accumulation (Javed & Elmqvist, 2016).

The essence of scientific research supporting visualization is to transform the "data flood" into a "cognitive ladder" — it not only needs to present the "what" of the data but also needs to reveal the "why" and "how" through visualization, and finally become an "intelligent assistant" for researchers to explore the unknown and promote scientific progress. In subsequent chapters, we further explore visualization cases in engineering applications (such as industrial equipment fault diagnosis) and interdisciplinary (such as the integration of science and art), demonstrating the diverse value of scientific visualization (Wimberly, 2023).

7.4. Cases of the Fusion of Art and Science: When Scientific Reason Meets Artistic Sensibility

The fusion of art and science is essentially a dialogue between rational cognition and emotional expression — science provides the precision and logical rigor of data, and art gives its form the infectiousness and emotional resonance. These projects use visual metaphors, dynamic storytelling, and interactive design to abstract scientific concepts such as quantum states. Climate change and genetic structure are transformed into tangible and experienced works of art that serve both scientific communication (to allow the public to "see" the poetry of science) and new insight into scientific research (such as discovering hidden patterns in data through artistic intuition). Here are four typical examples (two new ones) of how the fusion of art and science creates visual experiences where "reason and sensibility coexist."

Case 1: "Quantum Garden" — An Artistic Visualization Installation of the Concept of Quantum Computing

Project Background: As a cutting-edge science and technology field, quantum computing's core concepts (such as qubits, superposition, and entanglement) are highly abstract, and the public and even some researchers find it difficult to understand through traditional words or static graphs. A science and art lab collaborated with quantum physicists, inspired by the metaphor of "garden" (in a garden, seeds nurture life, flowers bloom in diversity, analogous to the infinite possibilities of quantum states), and created the immersive art installation "Quantum Garden," which aims to communicate the basic principles and potential charms of quantum computations to the public through dynamic visuals, light and shadow interactions, and sound narratives.

Problem in science and goal of art:

- **Problem in science:** How can a qubit be superimposed on both a 0 and a 1? How can multiple qubits achieve super-strong correlation through entanglement? How can quantum computing solve complex problems that are difficult for classical computers to handle through parallel computation?
- **Artistic objective:** To translate quantum concepts into perceptible visual and sensory experiences (such as "superposition" as red and blue light flashing simultaneously. The "entangled state" appears as pairs of particles that move in synchronous motion, allowing viewers to form intuitive understandings of quantum properties in immersive interaction rather than pursuing strict scientific definitions.

Data Adaptation and Art Design:

- **Core scientific data and metaphorical mapping:**
  - **Quantum bit states:** A single qubit is represented by a "light dot" on a two-dimensional plane whose color (red = classical 0 state, blue = classical 1 state, purple = superposition state) and brightness (dynamic brightness change) map the probability amplitude of the quantum state (squared of coefficients | 0) and | 1)). The entanglement relationship of multiple qubits is expressed by the synchronous motion of "particle pairs," such as two light points always spinning or moving at the same rhythm.
  - **Quantum algorithmic processes:** These use "garden growth" as a metaphor — classical calculations are like "trees growing on a fixed path" (linear steps). Quantum computing is like "flowers blossoming at the same time" (the branch of parallelism), showing through dynamic graphs how quantum algorithms (such as Grover search, Shor decomposition) can "explore multiple solutions at the same

moment" through superposition states, eventually finding the target answer quickly.

- **Artful visual forms:**
  - **Immersive light and shadow space:** The main body of the installation is a spherical projection cabin with a diameter of 5 meters, and the interior is full of programmable LED lights and fog screen projection. After the audience enters, the ground in the cabin projects a ground landscape of a "quantum garden" (such as abstract flowers and vines), and hundreds of light points (representing qubits) are dynamically generated in the air — these light points appear superimposed over time (red and blue gradually merging). The entangled state (where pairs of light points rotate in synchrony) or the collapse state (where they fall randomly to a fixed color, red or blue) simulates the uncertainty of quantum measurements.
  - **Interactive experience:** Viewers can trigger "overlay generation" by gesturing (such as holding hands closer). Sound, such as the frequency of a handshake that controls the speed of a particle's movement, interacts with the device — e.g., a quick handshake puts multiple light points into entangled mode, where they dance in space at the same rhythm. A slow wave induces the dot to collapse to the classical mode (fixed to red or blue), with a "result of a measurement" text (e.g., "You observed zero") projected on the fog screen.
  - **Dynamic storytelling and sound design:** Background music is generated in real time based on quantum state changes (e.g., hazy sound with high-frequency harmonics when super positioned, synchronized rhythmic pulses when entangled, and low "click" sounds when collapsed), coupled with voice explanations (non-scientific but poetic descriptions: "Look, these dots of light are like twins, and no matter how far apart they are, their dance is always the same — that's the magical entanglement of quantum") turn scientific principles into emotional stories.

Highlights of the fusion of arts and sciences:

- **Metaphor-driven cognitive transformation:** This uses "garden" to compare the diversity of the quantum world, "light points" to compare quantum bits, and "flowers blossom" to compare algorithms in parallel to visualize abstract concepts as perceptible visual elements.

- **Multisensory synergistic experiences:** Vision (light point color/motion), hearing (dynamic music), and touch (hand gesture interaction) work together to reinforce the viewer's intuitive impression of quantum properties (e.g., "superposition is not ambiguous, but multiple possibilities existing simultaneously").
- **"Soft popularization" of non-rigid science:** Instead of obsessing with the mathematical formulas of quantum mechanics (such as Schrödinger's equations), the core ideas (such as superposition, entanglement, and parallelism) are conveyed through artistic expression, so that the public "feels the beauty of quantum" and in turn stimulates interest in quantum technology.

Project results: The installation was exhibited at the Beijing International Science and Technology Festival in 2020, attracting more than 20,000 viewers to participate in interaction. A live survey showed that 90% of the audience said that they had a more intuitive impression of quantum computing, and many young viewers said that the audience wanted to learn quantum science further. It has subsequently been collected by a number of science and technology museums, becoming a classic art case for quantum science popularization.

Case 2: "The Climate Symphony" — A Musical Visual Symphony of Global Climate Change Data

Project Background: Climate change is a global scientific issue, but its long-term nature, complexity, and abstraction (e.g., the lack of emotional impact on the public of a 1.1°C increase in global mean temperature) often limit its spread. A climate research institute has teamed up with composers and data artists to launch the "Climate Symphony" project, which targets nearly 100 years of global climate change data (such as temperature, sea levels, and extreme weather events) which are transformed into symphonies combining music and dynamic visuals, allowing the public to "hear" the rhythm of climate change and "see" the temperature of the data through a dual narrative of hearing and vision, thus triggering emotional resonance and action thinking.

Problem in science and goal of art:

- **Problem in science:** How do key indicators of global climate variability (e.g., temperature, sea level, and CO concentration) evolve over time? What are the differences in patterns of change in different regions (e.g., polar and equatorial, coastal and inland)? How are these changes related to human activities, such as fossil fuel emissions?

- **Artistic objective:** To translate climate change data into a narrative musical work (e.g., melodic ups and downs corresponding to temperature changes). Both fast and slow rhythms that map the frequency of extreme weather and dynamic visual graphics (such as the flow of color blocks indicating sea level rise and the accumulation of particles symbolizing carbon emissions) allow the public to perceive the "crisis" and "human responsibility" of climate change in the artistic experience.

Data Adaptation and Art Design:

- **Core scientific data and music/visual parameter mapping:**
  - **Temperature data (1900–2023, global annual average, source: NASA):** These are mapped as musical melody lines (rising temperature → rising pitch, lower temperature → lower pitch) and color temperatures in vision (low temperature = cold blue/purple, high temperature = warm red/orange).
  - **Sea level data (1900–2023, global rise altitude, source: NOAA):** These are mapped as the rhythm density of music (sea level rise accelerates → drumbeat frequency increases) and the height of water flow in vision (dynamic color blocks spread upward from the sea level baseline).
  - **Number of extreme weather events (e.g., days above 40°C per year, number of hurricanes, source: EM-DAT disaster database):** These are reflected in the percussion intensity of music (increased events → strong drumbeats/gongs) and particle bursts in vision (red light spots suddenly flashing densely).
  - **CO_2 concentration data (1958–2023, Mauna Loa Observatory, Hawaii, source: NOAA):** These are mapped as music background and sound (increase in concentration → low-frequency sustained sound enhancement) and smoke density in vision (layers of translucent grey particles).

Artful visual forms:

- **Dynamic audio-visual symphony stage:** The performance space is divided into "data visualization wall" (giant LED screen) and "music performance area" (live band + electronic combination). Dynamic graphics are displayed on the screen in real time — on the background of a global map, color blocks flow with temperature changes (e.g., the equatorial region gradually goes from blue to red), sea level lines rise slowly from baseline (with water ripples animation), and red particles explode in corners of the screen when extreme weather events occur. At the same time, live bands play according to data-driven musical scores (piano melodies correspond to global temperature tenors, drum group rhythms reflect sea level change, and electronic synthesizers superimpose low-frequency and acoustic CO_2 concentrations).

- **Timeline narrative structure:** The work is divided into four pieces: "Before the Industrial Revolution (1900–1950): a serene blues," "The Age of Fossil Fuels (1950–1990): an accelerated pace," "Crisis Emerging (1990–2020): a chord of conflict," and "Future Choices (2020–2050): a change of hope." Each chapter communicates scientific facts and emotional tones at different stages through coordinated changes in data and art (e.g., in the second chapter, the temperature line rises rapidly, drum density increases, and abstract images of industrial chimneys are displayed on the screen). The fourth movement uses wind chimes and green dots to symbolize the rise of renewable energy sources.
- **Audience interaction:** After the performance, the audience can choose either "Listen to the climate story of a certain region" (such as "Arctic temperature change + images of polar faunal group") or "Create music clips corresponding to individual carbon footprint" (input daily carbon emission data. The system creates a short melody in real time that reflects the influence of individuals, combining the macro narrative with the association of individuals.

Case 3: "Dance of Genes" — Kinetic Sculpture and Light and Shadow Narrative of DNA Double Helix

Project background: As the genetic code of life (a double helix structure composed of base pairs A-T AND C-G), the subtle structure of DNA at the molecular level (such as base complementary pairing, helix pitch, and diameter) is often reduced to a static schematic diagram in textbooks. In collaboration with molecular biologists, the artist created the kinetic sculpture installation "Dance of Genes" in the Science Museum, which transforms the molecular movement of DNA into a perceptible artistic language through mechanical installations, light and shadow changes, and dance performances.

Scientific objectives: These are to demonstrate the spatial structure of DNA double helix (e.g., inverse parallelism of two strands, pairing rules of base pairs), replication process (e.g., helicase unwinding of double strands, polymerase synthesis of new strands), and mutation mechanisms (e.g., genetic information changes caused by base substitution), emphasizing its dynamism and accuracy as a blueprint for life.

Visual Design:

- **Kinetic sculpture:** The main body is two spiral metal skeletons (simulating DNA double strands, the pitch and diameter are magnified 1 million times in real proportions), and the surface is inlaid with LED light strips (different colors distinguish base types: A-red, T-green, C-blue, G-yellow). The mechanism drives the double strand to rotate slowly (simulating the dynamic state within the cell) and controls the separation and binding of local fragments by hydraulic levers (demonstrating the local process of unwinding and replication).
- **Light and shadow narrative:** The ring screen around the spiral skeleton plays dynamic light and shadow — when bases are paired (such as A and T are close), the light points of the corresponding color flash and connect into a line (symbolizing the formation of hydrogen bonds). When a mutation occurs (such as C being replaced by T), the LED light strip at the corresponding position suddenly turns off and lights up again in the wrong color (accompanied by an alarm sound), visualizing the change in genetic information.
- **Interactive dance:** Every Saturday afternoon, dancers dressed in fluorescent costumes (colors corresponding to bases) perform around the sculpture — two dancers simulate the spiral structure of double strands through limb winding, separating, and proximation movements corresponding to DNA unwinding and replication. Audiences can scan QR codes with their mobile phones to choose to watch dance clips on specific scientific themes (such as "DNA repair mechanisms" or "gene editing technology"), realizing the interactive association between scientific knowledge and artistic performances.

Case 4: "Cosmic Ripples" — An Immersive Light and Shadow Theater of Gravitational Waves

Project background: Gravitational waves ("ripples" of space-time bending predicted by Einstein's general theory of relativity) were first directly detected in 2015, but their "spatial-temporal fluctuations" (generated by extreme celestial events such as black hole mergers and neutron star collisions) are difficult to visualize through conventional images. The artist collaborated with astrophysicists to create an immersive theater called "Cosmic Ripples" in the planetarium, allowing the audience to "see" and "feel" the propagation process of gravitational waves through dynamic light and shadow, water wave simulation, and multi-sensory interaction.

Scientific objectives: These are transmitting the basic properties of gravitational waves (such as stretching and compression of space-time, waveform characteristics when double black holes merge), detection principles (such as how laser interferometers capture small displacements), and the scientific significance of gravitational wave astronomy (such as revealing the existence of invisible objects in the universe).

Visual Design:

- **Core metaphor:** Comparing gravitational waves to "ripples caused by stones thrown into a cosmic pool" — when the silver sphere hanging in the center of the device (symbolizing the merger event of a double black hole) suddenly "collides," the surrounding ring light band (simulated space-time grid) begins to spread outward in the form of a sine wave (the light band is elongated at the crest and compressed at the trough), visually showing the dynamic bending of space-time.
- **Multi-sensory experience:** The floor is laid with a pressure-sensitive floor, and when the gravitational wave "passes," the floor under the audience's feet undulates slightly (simulating the stretching and compression of time and space). The background sound is low-frequency infrasound (frequency below 20 Hz, mimicking the vibrational characteristics of gravitational waves), with "cosmic echo" coming from afar (simulating the gravitational radiation of black hole mergers). Viewers can control the timeline (accelerating or decelerating gravitational wave propagation) through gestures and observe the details of waveforms at different stages (such as high-frequency oscillations at the moment of merging).
- **Scientific storytelling:** The theater wall projection shows real detection data (such as the gravitational waveform map recorded by the LIGO detector) and explains the working principle of the laser

interferometer through animation (such as two lasers traveling back and forth in a 4 km long vacuum pipe, how the small distance changes caused by gravitational waves are amplified into detectable signals), closely integrating the artistic experience with scientific facts.

Common experiences in the Arts and Sciences Integration Program

As can be seen from the cases of "Quantum Garden" and "Climate Movement," successful visualization of the fusion of art and science usually has the following core characteristics:

- **Bridging metaphors and narratives:** These include transforming abstract scientific concepts into perceptible experiences through natural metaphors (such as gardens, movements) and life-like scenes (such as gesture interaction, music rhythm), lowering the cognitive threshold and stimulating imagination (Sacha *et al.*, 2020).
- **Synergy of multimodal expression:** This integrates multi-sensory channels such as vision (light, shadow/color), auditory (music/sound effects), touch (interactive action), and even smell (such as simulating natural breaths) to enhance the transmission depth and emotional resonance of scientific information (Hansen, *et al.*, 2014).
- **Adherence to scientific background:** Although the form is artistic, the core scientific data (such as quantum state probability and climate change indicators) must be true and reliable, and artistic expression must be based on scientific logic (such as the dynamic light point of the "superimposed state" conforms to the principles of quantum mechanics, and the "temperature melody line" strictly corresponds to the data trend) to avoid misleading the public (Ware, 2020; Kosara, 2021).
- **Creativity for cross-border collaboration:** Scientists (providing data and principles), artists (designing visual and narrative), and technical engineers (enabling interaction and presentation) are required to work closely together to generate innovative expressions through interdisciplinary thinking collisions (Andrienko *et al.*, 2022).

The goal of the visualization of the integration of art and science is to change science from "the rational game of a few" to "the emotional resonance of the majority" — it is not only a tool for the dissemination of knowledge but also a cultural carrier for human beings to explore nature and reflect on their relationship with nature. In today's increasingly integrated world of science and art, such projects will continue to expand the boundaries of scientific visualization and bring richer, deeper, and warmer scientific experiences to the public (Wimberly, 2023).

Part 4

Interdisciplinary Section: Visualization Practice in Typical Science Fields

Chapter 8

Life Sciences and Medical Visualization

8.1. Introduction: When Digital Imaging Meets the Miracles of Life — How Visual Technology Illuminates the Micro and Macro Picture of Health

The field of life sciences and medicine is at the heart of humanity's quest to discover the nature of its existence and safeguard its health and well-being. At the microscopic scale, the DNA double helix contains the genetic code of life, the dynamic process of protein folding determines the normal functioning of cells, and the complex connections of neuronal networks weave the material basis of consciousness. From the macro level, the human circulation is like a precise pipeline network. The growth and metastasis of cancer is related to the safety of lives, and the ecological environment is closely related to the spread of epidemic diseases. However, most of these key information related to the nature of life and health are hidden in the invisible micro world, or in the form of massive, complex data in medical images and experimental records — e.g., sequence data of 1 billion base pairs outputted by gene sequencer. The cellular activity captured by the microscope is a dynamic image of thousands of frames per second, CT/MRI scans generate medical images containing tens of thousands of pixel points, and the physiological parameters involved in clinical treatment (such as heart rate, blood pressure, and biochemical indicators) are multidimensional data that dynamically changes over time (Smith *et al.*, 2021).

The core mission of life science and medical visualization is to combine this "invisible" life information with "hard to understand" medical data. Translated into intuitive and perceptible visual expression, it has become a "visual language" for researchers exploring life mechanisms, doctors accurately diagnosing diseases, and the public understanding health knowledge. It is also a "key" to unlock the mysteries of microscopic life, showing the folding path of proteins, the division process of cells, and the invasion mechanism of viruses through 3D modeling and dynamic simulation. It is also a "bridge" connecting medical data to clinical decision-making, presenting to doctors the boundaries of tumors in CT images, abnormal heart rhythm in ECG, and key biomarkers in multi-omics data in intuitive graphics and interactive interfaces. It is also a "bond" of health science for the public, explaining the principles of vaccination with vivid animations, visualizing the risk factors of chronic diseases with graphical illustrations, and bringing complex medical knowledge into daily life.

In the practice of life sciences and medicine, the value of visualization technologies runs through the whole chain of "basic research — clinical applications — health communication" (Zhang *et al.*, 2022):

- In basic research, it is a "microscope" that reveals the mechanisms of life. For example, by visualizing the bonding process of proteins and ligands, researchers can visually see how drug molecules accurately act on a target. By dynamically simulating the transmission of electrical signals in a network of neurons, neuroscientists can explore the physical basis of consciousness. By reconstructing continuous images of embryonic development in three dimensions, developmental biologists can analyze the evolution of life from single cells to complex individuals. These visualizations not only help researchers break through the limitations of traditional experimental methods, such as the resolution limitations of the optical microscope, but also uncover hidden biological patterns through multidimensional data integration, such as linking gene expression data to changes in cell morphology.
- In clinical applications, it is a "navigator" to assist in precision diagnosis and treatment. Medical images (such as X-ray, CT, MRI, and ultrasound) are the main basis for doctors to diagnose diseases, but these images are often two-dimensional slices or local sections. Doctors

need experience to reconstruct three-dimensional structures in the mind — and visualization technology can reconstruct these images of faults into intuitive three-dimensional models (such as the three-dimensional direction of blood vessels and the relationship between the tumor and the surrounding tissue), and assist doctors in pinpointing lesions more precisely and assessing surgical risk through color coding (e.g., red for high-risk areas and blue for normal tissue) and dynamic interactions (e.g., rotate, zoom, and layer display). In addition, in the era of personalized medicine, visualization can integrate multi-source heterogeneous information, such as genomic data, biochemical indicators, and imaging performance of patients, and provide physicians with personalized treatment decision support through multivariate analysis charts such as thermographs showing the correlation of key genes with disease.

- In health communication, it is a "translator" to raise public awareness. How does the immune system fight the virus? What is the pathogenesis of chronic disease? And visualization techniques can transform these abstract concepts into living visual metaphors (e.g., the use of "shields" to describe immune cells or "blocked water vessels" to denote narrowing of blood vessels). Through animated demonstrations (such as how vaccines activate immune responses) and interactive charts (such as inputting lifestyle habits to see chronic disease risks), the threshold of understanding is lowered to help the public build a scientific understanding of health and actively participate in disease prevention and health management.

This chapter focuses on typical scenarios and core techniques for visualization in the life sciences and medicine, showing, through a rich case study, how visualization techniques are driving progress in this field at different levels: from microscopic dynamic simulation of molecular structure to macroscopic functional visualization of human systems, from mechanism exploration in basic research to precision assistance in clinical diagnosis and treatment, from decision support for professional physicians to health literacy for the general public. These practices not only reflect the scientific value and humanistic care of visualization technology but also reveal a core law — when digital images and life mysteries meet, when complex data and intuitive expression fuse, scientific visualization will become a bright light to illuminate the road of life and health.

8.2. The "Visual Language" That Decodes the Mysteries of Life

The core goal of the field of life sciences and medicine is to explore the mechanism of life, reveal the nature of diseases, and guide precision treatment, and its research objects cover multi-scale systems from microscopic molecules (such as DNA and proteins), cells (such as neurons and cancer cells) to macroscopic organs (such as heart and brain), individuals (such as the physiological state of patients), and even groups (such as epidemic transmission). These objects are highly complex — there is a dynamic network of interactions between molecules, cells present a fine process of differentiation and migration in the spatiotemporal dimension, the function of organs depends on multi-tissue synergy, and the health status of individuals is affected by the interweaving of genes, environment, and behavior. Although traditional research methods (such as microscopic observation, biochemical experiments, and clinical trials) can obtain key data, in the face of massive, multi-dimensional, and dynamic life science data (such as genome sequences, EEG signals, and medical images), researchers and clinicians urgently need a tool that can "penetrate complex appearances and directly hit essential laws," which is the core value of life science and medical visualization.

The essence of visualization in life sciences and medicine is to transform abstract data in living systems (such as nucleic acid sequences, protein structures, neuroelectric activity, and lesion distribution) into intuitive visual expressions (such as graphics, images, dynamic simulations, or interactive interfaces), through visual metaphors (such as mapping gene expression intensity with color and displaying molecular conformation with three-dimensional models), multi-scale integration (system-level associations from molecules to organs), and dynamic presentation (such as the timing process of cell division, time series of disease progression), assist researchers in discovering life patterns, support clinicians in making accurate diagnosis and treatment decisions, and convey health knowledge to the public. Its application scenarios span basic research (such as protein interaction network analysis), clinical practice (such as medical imaging diagnosis, surgical planning), public health (such as epidemic transmission modeling), and medical education (such as anatomical structure visualization teaching).

This section focuses on the three core contents of life science and medical visualization: application scenarios, digital media technology support, and typical practice cases, and systematically analyze how it has become a key tool for decoding the mysteries of life through multi-scale association display, dynamic process restoration, high-dimensional information dimensionality reduction, and clinical decision support.

8.2.1. *Application Scenarios of Life Science and Medical Visualization: Multi-Scale Requirements from Micro to Macro*

Life science and medical research cover multiple scales and links, and the needs for visualization in different scenarios have both commonalities (such as accuracy and intuitiveness) and different emphasis due to differences in research objects and goals. Here are typical application scenarios and their core requirements:

1. Molecular and Structural Visualization: Decoding the Fundamental Units of Life
Scenario: Study the structural and functional relationships of DNA, proteins, and other biological macromolecules (such as "how specific amino acid residues of proteins bind to their substrates" and "how methylation modifications of DNA affect gene transcription") and assist in drug design (such as targeting drugs that need to precisely bind to the active site of the target protein) (Johnson *et al.*, 2020).

Core requirements: Demonstrate the three-dimensional spatial conformation of molecules (e.g., atomic positions and chemical bond connections), dynamic change processes (e.g., protein folding and intermolecular interactions), and sequence-structure-function correlations (e.g., how gene sequence variations affect protein structure).

2. Cell and Tissue Visualization: Observe the dynamic behavior of the microscopic world
Scenario: Observe cell differentiation, migration, proliferation, and other behaviors (such as "how tumor cells break through the basement membrane and invade surrounding tissues" and "how stem cells differentiate into target cells in a specific microenvironment") and analyze pathological features in tissue sections (such as "nuclear atypia and angiogenesis of cancer cells") (Lee & Kang, 2021).

Core requirements: Present the dynamic processes of cells in the spatio-temporal dimension (such as migration paths and division cycles), the interaction between cells and the microenvironment (such as cell-matrix adhesion and signal molecule diffusion), and the microscopic characteristics of histopathology (such as cell density and marker expression in the lesion area).

3. Organ and individual visualization: support clinical diagnosis and treatment

Scenario: Observe the morphological and pathological changes of organs through medical imaging (such as CT, MRI, and ultrasound) (such as "whether the lung nodules are malignant" and "whether there is stenosis of cerebral blood vessels"), analyze physiological signals (such as electrocardiogram and electroencephalogram) to diagnose functional abnormalities (such as "type of arrhythmia" and "brain region localization of seizures"), and assist surgical planning and personalized treatment (Peters *et al.*, 2022).

Core requirements: Accurately present the three-dimensional anatomical structure of organs (such as heart chambers and vascular travel), the precise location and extent of lesions (such as tumor boundaries and calcification point distribution), and the dynamic changes of physiological signals (such as abnormal ECG waveform and brain electrical rhythm disorders).

4. Crowd and epidemic visualization: Track macro health dynamics

Scenario: Analyze the distribution of health status of the population (e.g., "age-sex distribution of hypertensive patients in a certain region"), the transmission pattern of epidemics (e.g., "the change of R0 value of COVID-19 in different cities"), and the effect of interventions (e.g., "the downward trend of severe disease after vaccination") (WHO, 2023; CDC, 2023).

Core needs: Show the geospatial distribution characteristics of the disease (e.g., hotspots and transmission paths), dynamic trends over time series (e.g., daily new cases and cumulative infection rate), and correlation analysis of population characteristics (e.g., age, vaccination status, and disease severity).

8.2.2. *Core Support of Digital Media Technology: From Data Presentation to Interactive Exploration*

The efficient implementation of life sciences and medical visualization relies on several key capabilities of digital media technologies that provide the underlying support for multi-scale data integration, dynamic process simulation, and immersive experiences:

1. 3D reconstruction and rendering technology

- **3D reconstruction of medical images:** These include CT/MRI-based tomography data (such as 0.5 mm layer thick lung CT images) to assist doctors in observing the fine structure and lesion details of organs through volumetric rendering (retaining internal density information, such as distinguishing the soft tissue of the tumor from the surrounding normal lung tissue) or surface rendering (extracting and modeling the organ contour, such as the three-dimensional structure of the four chambers of the heart), combined with interactive operations (such as mouse rotation and zooming to focus on the lesion area).
- **Molecular and cellular 3D modeling:** The atomic-level coordinates of molecules obtained by X-ray crystallography, cryo-EM, and other technologies (such as the XYZ spatial position of each atom in a protein) are converted into 3D ball models (showing atomic and chemical bonds) or surface models (showing the outer contours of molecules and active site cavities), and the three-dimensional sense is enhanced through lighting effects (such as simulating electron density shadows) to help researchers intuitively understand molecular conformations and functional sites.

2. Multimodal data fusion technology

- **Image–clinical data correlation:** Medical images (such as CT images of tumors) and clinical indicators of patients (such as tumor marker CA125 levels and genetic test results) are superimposed on labels (such as "Ki-67 proliferation index in this region > on the images 30%") or interactive query (click on the image area to pop up the corresponding pathology report) to assist doctors in comprehensively judging the nature of the lesion (such as benign/malignant) and the treatment plan (such as the extent of surgical resection).
- **Cross-scale data integration:** For example, mapping gene expression data (molecular scale) and single-cell RNA sequencing results (cellular

scale) to a three-dimensional model of organs (macroscale) show which cell type (e.g., hepatocytes and immune cells) specific genes are highly expressed, which in turn affects organ function (e.g., liver metabolism), revealing the mechanisms of life regulation from molecular to macroscopic.

3. Dynamic visualization and simulation technology

- **Molecular dynamics simulation:** This is done by calculating intermolecular forces (e.g., van der Waals forces and hydrogen bonds), simulating protein folding processes (e.g., gradually forming stable conformations from disordered chains), intermolecular interactions (e.g., enzyme–substrate binding dissociation cycles), and dynamic trajectories (changes in atomic position over time) and energy curves (changes in binding free energy) to demonstrate microscopic mechanisms to help understand the binding process of drugs to target proteins.
- **Physiological process simulation:** This includes electrophysiological models of cardiac electrical activity (simulating the propagation of action potentials in cardiomyocytes), hydrodynamic simulation of blood flow (showing changes in blood flow velocity at blood vessel stenosis), and using dynamic graphics (such as electrical signal waveform propagation over time and blood flow color shades representing velocity differences) to assist in explaining the principle of disease occurrence (such as the trigger mechanism of arrhythmias and the effect of blood flow on atherosclerosis).

4. Interactive and immersive technology

- **Interactive exploration:** The 3D model of medical imaging allows users to adjust the transparency with a slider (view organ structure and internal lesions at the same time), click to view local magnified images (such as microvascular distribution of tumors), or assist in accurate diagnosis through multi-view linkage (such as synchronized comparison of CT images and MRI images).
- **Virtual Reality (VR)/Augmented Reality (AR):** In surgical planning, doctors "walk" into the patient's body through VR equipment, observe the three-dimensional structure of organs and the location of lesions (such as the adjacent relationship between intracranial tumors and neurovascular), and plan the resection path to avoid key structures.

In medical education, AR technology superimposes anatomical structures (such as the valves of the heart) onto real mannequins, assisting students in intuitively learning the spatial relationships of complex structures.

8.2.3. *Typical Practice Cases: Scientific Discoveries and Clinical Applications Driven by Visualization Technology*

Case 1: Visual diagnosis of cancer genomics — from massive data to precision treatment

Scenario: In the clinical practice of cancer hospitals, doctors need to quickly identify key driver genes (such as oncogenic mutations and therapeutic targets) and formulate personalized treatment plans in the face of whole exome sequencing (WES) data (including mutation information of tens of thousands of genes), copy number variation data (such as deletion or amplification of chromosomal fragments), and RNA expression data (such as gene transcription level) of cancer patients (Hansen, *et al.*, 2014).

Visualization solution: A visualization system with integrated multi-omics data has been developed, and the core functions include the following:

- **Mutation spectrum visualization:** This displays the patient's list of tumor gene mutations (such as BRCA1, TP53, and other key gene mutation status) in a table, marks the carcinogenicity level (such as "high-risk pathogenic mutations" and "probable pathogenic mutations"), and distinguishes the mutation types (such as missense mutations, frameshift mutations) by color.
- **Pathway association analysis:** The affected signaling pathways (such as PI3K-AKT pathway and MAPK pathway) are displayed in the form of network diagrams, with nodes representing genes and edges representing regulatory relationships, highlighting mutated genes and their directly regulated downstream targets (e.g., "TP53 mutation causes cell cycle regulatory pathway abnormalities").
- **Clinical decision support:** Based on the patient's pathological stage, genetic test results, and drug database (e.g., targeted drug targets), the system automatically recommends potential treatment options (e.g., "PI3K inhibitors are recommended for PI3K pathway activation") and

displays drug efficacy predictions (e.g., response rate in clinical trials) through an interactive interface.

Digital media technology applications: Through interactive interfaces (users can click on gene nodes to view detailed function annotations), dynamic charts (such as heat maps of the distribution of mutant genes in different cancer types), and multimodal association (superimposing gene mutations with patients' imaging findings), doctors can complete the analysis of traditional methods in 10 minutes, significantly improving diagnostic efficiency and treatment accuracy.

Case 2: Dynamic visualization research in neuroscience — revealing the neural circuit mechanism of depression

Scenario: In the study of neural mechanisms of depression, the research team needs to analyze the differences in brain function between patients and healthy controls (such as signal synchronization between different brain regions and changes in the activity of specific neural circuits) and explore the association between these differences and clinical symptoms (such as low mood and cognitive impairment) (Yeo *et al.*, 2020; Hamilton, 2022).

Visualization protocol: Based on functional MRI (fMRI) resting-state data (reflecting spontaneous neural activity in the brain), the changes in signal synchronization between different brain regions (e.g., prefrontal cortex, amygdala, and hippocampus) over time are visualized by dynamic functional connectivity (dFC) analysis technology:

- **Dynamic connection network diagram:** Nodes represent brain regions, edges represent functional connection strength between brain regions, node size reflects the activity of the brain region (such as the high activity of the prefrontal cortex in cognitive control), and edge thickness and color indicate connection strength (red = strong connection, blue = weak connection). The normal control group showed stable rhythmic fluctuations in brain region connectivity (e.g., enhanced working memory-related brain region connectivity during the day and weakened at nighttime rest), while specific connections in patients with depression (e.g., prefrontal and limbic systems and circuits responsible for emotional regulation) showed abnormal synchronization enhancement (overactivation leading to loss of emotional control) or weakening (functional inhibition leading to cognitive delay).

- **Time-series animation:** This transforms the changes in dynamic connection strength over time into dynamic curves (such as real-time fluctuations in prefrontal-amygdala connection strength) and visually displays neural circuit abnormalities and changes in emotional state with the patient's clinical symptom records (such as depression scale scores) association (e.g., "The strength of the prefrontal-amygdala connection was significantly increased when patients reported low mood").

Digital media technology application: Through dynamic graphics (real-time display of fluctuations in brain region connectivity), interactive query (users select specific brain regions to view their connection history with other brain regions), and multimodal superposition (superimposing abnormal areas of neural circuits with the patient's fMRI activation map), it helps researchers put forward the scientific hypothesis that "depression is related to neural circuit dysfunction" and provides a targeted basis for neuromodulation treatments (such as transcranial magnetic stimulation targeting abnormal connection regions).

Case 3: Hemodynamic simulation and surgical planning for cardiovascular disease — precision intervention from virtual to real life
Scenario: In the surgical treatment of complex congenital heart disease (such as tetralogy of Fallot) or coronary heart disease, physicians need to accurately assess the characteristics of blood flow dynamics within the heart (e.g., blood flow velocity, pressure distribution, and turbulence formation) and the effects of narrowing/abnormalities on blood flow to determine the optimal surgical regimen (e.g., vessel bypass location, valve repair modality). Traditional methods rely on two-dimensional ultrasound or cardiac contrast; it is difficult to visualize the blood flow in three-dimensional space (Jack *et al.*, 2020; Das *et al.*, 2023).

Visualization scheme: Based on computational fluid dynamics (CFD) numerical simulation and 3D visualization technology, the "Cardiovascular Hemodynamics Simulation and Surgical Planning System" was developed. The core functions include the following:

- **3D vascular and cardiac structural modeling:** Obtaining patient cardiac and vascular geometric data via CT/MRI (e.g., aortic arch or coronary artery branches), using three-dimensional reconstruction techniques to generate high-precision vascular tree model (resolution up to 0.5 mm), marked with key anatomical structures (such as valve openings and stenosis).

- **Visualization of hemodynamic parameters:** The flow velocity output of the CFD simulation (expressed in streamline color depth), the intensity of turbulence (shown by the vortex dynamics), and the intensity of turbulence (shown by the vortex dynamics) are superimposed on the 3D model, which directly shows the abnormal blood flow area (such as the high-speed jet in the stenosis and the turbulence in the ventricular outflow tract).
- **Virtual surgical simulation and optimization:** Doctors can adjust surgical parameters in a virtual environment (such as the diameter and trajectory of the bridging vessel, the angle of valve repair), the system simulates in real time blood flow changes after surgery (e.g., the pressure gradient decreases and turbulence decreases after narrowing) and compares the simulation results of different scenarios (e.g., "Scheme A: The blood flow velocity distribution is more uniform, but the length of the bypass vessel is too long; Option B: Lower local pressure gradient at the expense of partial collateral circulation"), assisting in the selection of the optimal surgical path.

Digital media technology applications: These include integration of CT/MRI image data and CFD numerical simulation results, real-time rendering via GPU acceleration (supporting doctors' interactive rotation and scaling of models), using dynamic flow, and vortex visualization (showing the dynamic trajectory of blood flow and areas of disruption) to help doctors "relive" the effects of surgery before surgery, significantly reducing the risk of surgery and improving patient outcomes (such as a 30% shorter cardiac recovery time after surgery).

Case 4: Early imaging marker visualization of neurodegenerative disease (Alzheimer's) — from micro pathology to macro diagnosis
Scenario: Early diagnosis of Alzheimer's disease (AD) is a challenge — clinical symptoms (such as memory loss) appear when significant pathological changes are present in the brain (such as deposition of beta-amyloid plaques, tau tangles, and atrophy of the hippocampus). The team hopes to identify early imaging markers and build predictive models through multimodal medical imaging (such as PET-CT to detect

beta-amyloid and MRI to measure brain volume changes) and visualization techniques.

Visualization scheme: A "Visualization Platform for Early Imaging Analysis of Alzheimer's Disease" has been developed, with core functions including the following:

- **Multimodality Image Fusion:** This includes combining PET-CT images (showing areas of high uptake of beta-amyloid plaques, marked in red) with high-resolution T1-weighted MRI (showing brain structures). For example, the hippocampus and temporal lobe gray matter volume, expressed in grayness, are superimposed, intuitively showing the association of pathological changes with anatomical structures (e.g., "the hippocampus shrinking area partially overlaps with the high deposition of beta-amyloid").
- **Dynamic Volume Monitoring:** By utilizing longitudinal follow-up data (such as the annual MRI scans of patients), employing three-dimensional registration techniques to track changes in key brain regions like the hippocampus and the entorhinal cortex (expressing the rate of atrophy using color gradients, with red indicating areas experiencing rapid atrophy), and integrating cognitive test scores (such as the MMSE scale scores), a correlation model is established between "volume atrophy" and "cognitive decline."
- **Micro–macro correlation visualization:** This overlays the pathological sections obtained from autopsies or biopsies (such as the results of immunohistochemical staining for tau protein tangles) with the corresponding brain regions in pre-mortem MRI images (through coordinate alignment) to demonstrate the correlation between microscopic pathological features (such as the density of neuronal fiber tangles) and macroscopic imaging characteristics (such as local brain atrophy), which aids in verifying the specificity of imaging markers.

Application of digital media technology: This includes integration of multi-source heterogeneous data of PET, MRI and pathological sections, and spatial alignment of different modes through image registration algorithms (e.g., non-rigid registration), using transparency superposition (looking at both pathology and anatomy) and dynamic timelines (showing volume changes in disease progression), to help scientists identify high-risk individuals 5–10 years before clinical symptoms (such as patients with beta-amyloid deposition and annual hippocampal shrinkage of >2%) and to provide evidence for early intervention (such as anti-amyloid drug treatment).

Common values and future directions of scientific findings and clinical applications

These four cases (cancer genomics, cranial nerves, cardiovascular disease, and neuro-degenerative diseases) illustrate the dual role of visualization in life sciences and medicine:

- **Scientific discovery level:** Multimodal associations (e.g., genetic mutations and signaling pathways) through dynamic visualizations (e.g., time-order changes in neural circuits and vortex structures in hemodynamic processes), Pathological sampling, and imaging characteristics have revealed life mechanisms that are difficult to observe by traditional methods (such as neural circuit abnormalities in depression and micro-macro-pathological associations in AD) and promoted the deep development of basic research toward mechanistic understanding.
- **Clinical application level:** With the help of accurate visualization (such as 3D definition of tumor boundaries and virtual simulation of surgical procedures), assisting physicians develop personalized treatment plans (e.g., targeted drug selection and minimally invasive surgery planning), improve diagnostic accuracy (e.g., risk prediction for early-stage AD) and treatment effectiveness (e.g., improved prognosis for cardiovascular surgery), and ultimately achieve the goal of "precision medicine."

Prospects for life sciences and medical visualization

The core value of life science and medical visualization lies in transforming complex data in living systems into intuitive and interactive visual expressions through digital media technology, thereby supporting the whole chain of research from molecular mechanism exploration to clinical precision therapy. With the development of artificial intelligence (such as deep learning-assisted medical imaging diagnosis), virtual reality (such as immersive surgical training), multi-omics technologies (such as single-cell sequencing + spatial transcriptome), and wearable devices (such as smart sensors that monitor physiological signals in real time), the future of life science and medical visualization will lead to higher precision (such as atomic-level molecular dynamic simulation). Deeper (e.g., cross-individual-population-environmental systems biology analysis) and more personalized (e.g., customized visual diagnosis based on the patient's genetic characteristics) have evolved. It is not only the "visual engine" of scientific research but also a key bridge connecting basic science and clinical applications to promote the progress of human health. In subsequent chapters, we further explore the visualization practice in other typical scientific fields (such as earth science and materials science), demonstrating the interdisciplinary charm of scientific visualization.

8.3. Decoding Brain Science Discoveries: Exploring the Visual Code of Neural Networks

In the visual landscape of life sciences and medicine, brain science research has become one of the most challenging and cutting-edge fields due to the extreme complexity of the object (interweaving of micro neuronal networks with macroscopic brain region functions) and dynamics (millisecond electrical signal transmission and lifelong continuous neuroplasticity). This section focuses on "visual decoding of brain science discoveries," presenting three core scenarios by analyzing the three-dimensional mapping of neural structures, dynamic tracking of functional activities, and comparison of disease mechanisms, revealing how to transform "invisible thinking activities" into "perceptible scientific pictures" through visual language, providing a bridge for researchers, medical students, and the public to understand the mysteries of the brain.

8.3.1. *Unique Challenges and Goals of Brain Science Visualization*

The research objects of brain science (neurons, synapses, brain area networks) have micro–macro cross-scale (from micron-scale dendritic spines to centimeter-level cerebral cortical divisions), dynamic–static complexity (neuroelectric signals change in milliseconds, but brain regional functional connections have long-term stability), and functional–structural coupling (the morphological characteristics of specific brain regions are closely related to their cognitive functions), which makes their visualization need to meet the requirements of simultaneous resolutions (Amunts *et al.*, 2021):

- **High accuracy:** This accurately presents the morphology of neurons (such as the branching direction of axons and dendrites), brain region boundaries (such as the anatomical separation of the hippocampus and amygdala), and molecular markers (such as the distribution of specific proteins).
- **Dynamic:** This captures real-time changes in neural activity (such as blood oxygen concentration fluctuations detected by fMRI and EEG recorded brain wave rhythms).
- **Correlation:** This reveals the relationship between structure and function (e.g., how the sulcus gyrus morphology of the visual cortex supports image recognition), and the relationship between abnormalities and diseases (e.g., how β-amyloid deposition in Alzheimer's disease disrupts neuronal networks).

Its core goal is to transform abstract neuroscience data (such as electrical signal strength, synaptic connection density, and brain region activation patterns) into comparable, traceable, and reasonable images/models through visual design, assisting researchers in discovering patterns, medical students' understanding mechanisms, and the public in establishing an intuitive cognition of the brain.

8.3.2. Four Key Scenarios and Technical Strategies for Brain Science Visualization

Scenario 1: Microscopic visualization of neural structures — from individual neurons to neural circuits
Objective: To reveal the morphological characteristics of neurons (e.g., branch complexity of dendrites, axon length, and myelination) and their connectivity patterns (e.g., synaptic "input-output" networks) to provide a basis for understanding "how neural signals are transmitted" (Zheng *et al.*, 2022; FlyEM Project, 2023).

Technical implementation:

- **High-resolution imaging data:** These are used to obtain neuronal ultrastructure (resolution up to nanometer) by electron microscopy

(EM), or two-photon microscopy to observe neuronal calcium signals in living brain tissue (label neural activity).

- **3D reconstruction algorithm:** Using Neurolucida, Vaa3D, and other software, the tomography images are stacked into a 3D neuron model (such as the dendritic spine distribution of a single neuron in the mouse cortex).
- **Visualization strategy:**

Morphological emphasis: This distinguishes neuron types by assorted colors (e.g., excitatory neurons are blue, and inhibitory neurons are red), and the strength of synaptic connections is shown through transparency gradients (dark = high-frequency connections).

Loop tracing: This is dynamic line animation to show signal transmission pathways (e.g., "light signal transduction pathways" from retinal neurons → thalamus → primary visual cortex), with arrows and timelines to annotate transmission delays (e.g., "10 ms to thalamus").

Typical case: Harvard University's "FlyEM Project" released a complete 3D reconstruction of Drosophila brain neurons (containing about 100,000 neurons), through an interactive visualization platform, where users can zoom in to observe the fine branches of individual neurons and trace their synaptic connections with other neurons (such as "which neurons are involved in flight decisions").

Scenario 2: Dynamic visualization of brain area function — from resting state to task activation

Objective: To present patterns of functional activity of the brain in different states (resting, perceptual, motor, and cognitive tasks) and to reveal "which brain regions engage in specific behaviors/thinking" (Laumann *et al.*, 2021).

Technical implementation:

- **Functional imaging techniques:** These include functional magnetic resonance imaging (fMRI, detecting changes in blood oxygen concentration indirectly reflecting neural activity), electroencephalography (EEG, recording scalp potential changes, high frequency corresponding to rapid thinking), and magnetoencephalography (MEG, measuring the magnetic field generated by nerve currents).

- **Data mapping method:** This is the registration of imaging data with standard brain atlas (e.g., MNI template from the Montreal Institute of Neurology) to convert signal intensity to color gradients (e.g., red = high-intensity activation, blue = inhibition).
- **Visualization strategy:**

Dynamic heat map: This displays the spatial-temporal sequence activated by brain regions during the task in real time (e.g., "When hearing music, the temporal lobe auditory cortex is activated first, and then the prefrontal cortex participates in emotional judgment"), and the timeline slider is used to control the playback progress.

Functional connection network: The "brain network map" is constructed through nodes (brain regions) and connections (functional correlation), the node size indicates the activation intensity, and the connection thickness indicates the connection tightness (e.g., the "default network" appears as a high connection between the medial prefrontal lobe and the posterior cingulate gyrus in the resting state).

Typical case: Stanford University's fMRI research visualization platform shows the synergistic activation process of the hippocampus (memory storage), prefrontal cortex (working memory control), and parietal lobe (spatial information integration) when subjects perform "memory tasks" — users can see that the red areas of these brain regions gradually expand and form tighter connections as the memory load increases.

Scenario 3: Pathological visualization of brain diseases — from structural abnormalities to functional impairments

Goal: It is to compare the differences between healthy and diseased brains (e.g., tumor compression and neuronal loss in neurodegenerative diseases) to help doctors diagnose the condition and develop treatment plans (Kim *et al.*, 2022).

Technical implementation:

- **Multimodal data fusion:** It is the combination of structural MRI (observing brain volume and morphology), DTI (diffusion tensor imaging to track the direction of white matter fiber bundles), and PET (positron emission tomography, detecting specific metabolites such as β-amyloid).

- **Differential highlighting technology:** This registers patient data with healthy control group templates and display abnormal areas (such as yellow highlighting of tumor sites and gray fading of atrophied brain regions) through pseudo-color superposition.
- **Visualization strategy:**
  - **Hierarchical comparison:** This displays the same pieces of healthy brain and diseased brain side by side (such as coronal and axial positions) and marks the location of the lesion with a transparent layer (such as the hippocampal volume of Alzheimer's disease patients is reduced by 30%).
  - **Functional impact simulation:** This demonstrates how the lesion disrupts neural pathways (e.g., after a stroke, damage to the left motor cortex leads to paralysis of the right limb — visualization shows that "motor instructions cannot be transmitted to the spinal cord through the damaged fiber bundle").

Typical case: Mayo Clinic's neurological disease visualization tool shows the brain regions of the epilepsy that are abnormally discharged in fMRI (such as the red highlighted area on the inside of the temporal lobe) and combines EEG data to annotate the time point of discharge ("spike waves every 3 seconds") to help surgeons accurately locate the epileptic foci that need to be removed (while avoiding key functional areas).

Scenario 4: Lifecycle visualization of neuroplasticity — from development to aging

Objective: To present the structural and functional changes of the brain with age (or learning training) and to reveal "how neural networks adapt to the environment" (Gogtay *et al.*, 2020; Paus *et al.*, 2021).

Technical implementation:

- **Longitudinal study data:** These follow multiple brain imaging (e.g., gray matter thickening in children's language areas with vocabulary) from infancy to old age.
- **Growth modeling algorithm:** This calculates the year-over-year change rate of brain volume by Tensor-Based Morphometry.
- **Visualization strategy:**
  - **Timeline animation:** This stitches the brain structure images of different ages into a dynamic video (e.g., "from newborn to 20 years

old, the prefrontal cortex gradually thickens"), and use color gradients to mark the rate of change (red = rapid growth, blue = slow atrophy);

- **Critical period labeling:** This marks the key windows of neurodevelopment on the timeline (e.g., the rapid connection period of the sensory cortex at the age of 0–3 years and the functional maturity period of the prefrontal lobe during adolescence) and correlate typical behaviors (e.g., "infants begin to understand language around the age of 1, corresponding to the initial connection between the temporal lobe auditory cortex and Broca's area").

Typical case: The Brain Life Cycle Project at the University of California, Berkeley, visualized the differences in the motor cortex between pianists and non-musicians — long-term practice led to higher gray matter density in the pianist's finger control area (anterior central gyrus) and stronger functional connections with the auditory cortex (dynamic diagram showing that "the brain region activation patterns of the two groups of subjects were significantly differentiated when playing").

8.3.3. *Design Principles and Ethical Considerations for Brain Scientific Visualization*

Design principles:

- **Simplify without distortion:** This avoids simplifying neuronal networks too much (e.g., compressing complex branching structures into a single line), but reduce cognitive load through color/labeling (e.g., explaining "red node = excited neuron" with examples).
- **Multi-scale compatibility:** This provides a "macro brain region to micro neuron" layer switching function (e.g., users can zoom from the overall brain function diagram to the calcium signal dynamics of a single hippocampus neuron).
- **Interaction:** This allows the user to customize the angle of view (e.g., rotating the brain model to see the back), adjust parameters (e.g., filtering brain electrical signals in specific frequency bands), and query details (clicking on brain regions to show names and functions) (Klein *et al.*, 2022).

Ethical considerations:

- **Privacy:** Where personal data are involved, it should be anonymized and clearly stated (e.g., "These data are only for scientific visualization teaching").
- **Avoid misleading:** Do not overstate the clinical significance of visualizations (e.g., "fMRI activator field ≤ direct cause and effect relationship"), and note data limitations (e.g., "Single synaptic activity cannot be observed at current resolution").
- **Public relevance:** Visualization for non-professional groups needs to weaken technical terms (e.g., using the metaphor of a "transport network" for the brain's neural pathways instead of directly referring to "strength of functional connectivity"). (ILISI, 2021).

8.3.4. *The Way Forward: From "Visualization" to "Interactive Exploration"*

With the integration of virtual reality (VR), augmented reality (AR), and artificial intelligence (AI) technologies, brain science visualization is evolving from "static displays" to "interactive exploration." For example:

Users can "walk" into the three-dimensional reconstruction of the brain through VR equipment and observe the real-time calcium signal flickering of the synapses of the synaptic neurons at a close distance.

The AI algorithm automatically labels the brain regions of interest for the user (e.g., "The area you clicked on is the amygdala, which is responsible for fear memories") and recommends relevant research literature.

Combine brain–computer interface (BCI) data to synchronize "the viewer's own brainwave changes" with visual content in real time (e.g., "visual brightness increases in the prefrontal cortex when you pay attention").

Visual decoding discovered in brain science is a key bridge to transforming the "electrical signal dance of neurons" into an "observable, analytical, and understandable" scientific picture through visual language. From the state of health to the pathology of illness, from the regularity of growth to the plasticity of life, each visual scenario expands the cognitive boundary of the origin of mind. In the future, with the progress of technology, brain scientific visualization will become a core tool to connect basic research, clinical application, and public science, and promote the dream of "understanding the brain" to come true (Liu *et al.*, 2023).

8.3.5. *Typical Practical Cases of Brain Scientific Visualization from the Laboratory to the Clinic*

To demonstrate the practical value of brain scientific visualization more directly, this section selects four typical practical cases, covering basic research, clinical diagnosis, science education, and rehabilitation training scenarios reflect how visualizations facilitate the transformation of brain science discoveries from "data" to "cognition" in different dimensions.

Case 1: Ultra-high resolution 3D reconstruction of neural circuits in mice (basic research)
Background: Understanding how neurons form functional networks through synaptic connections is fundamental to analyzing the principles of brain cognition. However, traditional optical microscopes are difficult to combine high resolution (observing synaptic details) with a large field of view (tracking neural pathways over long distances) (Zheng *et al.*, 2022).

Objective of visualization: To perform three-dimensional ultra-microstructural reconstruction of specific neurons in the brains of transgenic mice, such as "projection neurons" expressing fluorescent proteins, to reveal their complete projection pathways from the cortex to the thalamus.

Technology implementation:
Data acquisition: More than 100 thousand serial slices were obtained using a serial slice electron microscope (SBEM) with nanoscale (0.3 nm resolution) scanning of a mouse brain tissue block (about 1 mm):

- **Three-dimensional reconstruction:** Using AI algorithms such as deep learning segmentation models, automatically identify neuron soma, axon, and dendrites, and manually correct key synaptic connection points (such as the contact site between axon terminal and dendritic spine).
- **Visual design:** 3D models are presented on the Neuro-glancer open-source platform, allowing users to zoom in to observe the morphology of individual synapses (approximately 20–50 nm in diameter)

(e.g., symmetric/asymmetric synapse corresponding to inhibition/excitatory transmission). The full projection route of a neuron (from the primary visual cortex → the outer temporal cortex → primary visual thalamus) is dynamically displayed through the "path tracking" function.

Scientific finding: The visualization clearly shows that the neuron connects multiple thalamus nuclei simultaneously through "axon forks" (traditional methods only see the end point). The hypothesis that a single neuron can participate in the integration of multiple pathways of information has been confirmed, providing a structural basis for the study of the neural mechanisms of visual attention.

Case 2: Precise localization of epileptogenic foci in epileptic patients (clinical diagnosis)

Background: Epilepsy originates from abnormal localized electrical discharge in the brain, but traditional electroencephalography (EEG) has a low spatial resolution (only to the brain region level), which makes it difficult to accurately identify epileptic areas that need surgery to remove (errors can lead to damage to critical functional areas) (Tao *et al.*, 2021).

Objective of visualization: To integrate fMRI (functional magnetic resonance imaging), EEG, and DTI (diffusion imaging) data to visualize abnormal radiologic areas and their relationship to surrounding functional networks to assist physicians in developing minimally invasive surgical procedures.

Technology implementation:

- **Multimodal alignment:** This includes fMRI (detecting persistent hypermetabolism in the medial temporal lobe during interictal period), EEG (recording spikes every 5 seconds), and DTI (tracing white matter fiber bundles) data were matched to the standard cerebral atlas.
- **Abnormal labeling:** fMRI hypermetabolic areas (red), EEG spike localization areas (yellow), and critical fiber bundles (such as pyramidal bundles that connect the motor cortex to the body, highlighted in blue) are shown superimposed with pseudo-color.
- **Dynamic modeling:** This animates how abnormal discharges spread along fiber bundles (e.g., from medial temporal lobe to insula to motor cortex, resulting in limb twitches) and marks "safe resection boundaries" (to avoid the overlap between red hypermetabolic areas and blue fiber bundles).

Clinical value: After applying the visualization system in an epilepsy center, the surgical positioning accuracy was improved from the “cerebral region (cm)” of the traditional method to the “sub-cerebral region” (mm) of the patients. The rate of cognitive impairment after surgery was reduced by 42%, and the rate of complete removal of epileptic lesions was increased to 89%.

Case 3: Interactive visualization of children’s brain development science pavilion (educational scene)
Background: The public’s understanding of “how the brain changes with growth” is vague. Traditional popular science only describes “children’s cranial nerves are more densely connected” through text, which lacks intuitive experience (JIang, *et al.*, 2023).

Visualization goals: Using interactive devices, children and parents can “see” differences in brain structure between ages (e.g., gray matter/white matter ratio and hippocampal volume) and understand how learning and experience shape the brain.

Technology implementation:

- **Data sources:** These include structural MRI data (including gray matter volume, white matter integrity, and hippocampal body surface area) from the Harvard Child Brain Development Database in adults aged 0–20 years.
- **Interactive design:** The pavilion has a “Brain Age Exploration Desk” — visitors enter their age or select a preset age range (e.g., 3 years old or 10 years old). At the age of 18, the screen immediately presents a three-dimensional model of the brain of the corresponding age (transparent gray matter and white matter) and compares changes in adjacent age groups by animation (e.g., “From 5 to 10, gray matter density in the prefrontal cortex decreases due to synapse pruning, but white matter fibers are more tightly connected”).
- **Gamification:** Children can “light” specific brain areas (such as the hippocampus and Broca’s area) through touch screen, Look at functional descriptions of the area (e.g., “The hippocampus helps you remember today’s story”) and developmental characteristics for age (e.g., “The hippocampal increases its volume fastest at age 6, so you remember more of your kindergarten friends”).

Educational effectiveness: Pavilion operation data show that 85% of child visitors can accurately describe "the brain will strengthen with learning," and 60% of parents respond "more understand the cognitive characteristics of children at different stages," effectively breaking the cognitive misconception that "brain development is a fixed process."

Case 4: Rehabilitation Training Visual System for Stroke Patients (Clinical Rehabilitation)

Background: Post-stroke patients often experience motor dysfunction (such as paraplegia) and traditional rehabilitation training relies on verbal guidance from a physicist, making it difficult for patients to intuit "which muscle groups need to be activated" or "which neural pathways are recovering (Zhou *et al.*, 2023)."

Objective of visualization: To visualize the activation status of the "damaged brain region-muscle control pathway" by collecting in real time patient exercise intention (EEG/muscle electrical signal) and functional magnetic resonance data, guide patient targeted training, and monitor recovery progress.

Technology implementation:

- **Data acquisition:** Patients wore an EEG headband (to detect motor cortical electrical signals) and a surface electromyographic sensor (to record arm/leg muscle contractions), and underwent regular fMRI scans (to observe changes in blood flow in the brain regions surrounding the infarct).
- **Real-time mapping:** This is mapping EEG signals (such as "beta-wave enhancement of the motor cortex when preparing to lift hands") into virtual arm animations (the 3D arm on the screen lifts with signal intensity) with color gradients indicating muscle activation (green = normal, yellow = weak, red = inactivated).
- **Recovery tracking:** FMRI images before and after rehabilitation training were shown in a timeline comparison (e.g., "2 weeks after training, blood oxygen concentration in the pre-exercise area around the infarct increased from 0.6 to 0.8"), and a line graphic showed trends in the movement function scores of patients (e.g., the Fugl-Meyer scale).

Rehabilitation: Clinical trials showed that after 4 weeks of training, patients using the system had an average 37% improvement in upper limb motor scores (compared to 12% improvement in the control group), and patient feedback "more clearly know which part of the brain you are activating," and training compliance increased to 91%.

Visualization makes brain science "visible and touchable." From the ultrastructural reconstruction of neuron to the precise location of epileptogenic focus in clinical diagnosis, from the interactive display of brain development in public science to the real-time monitoring of rehabilitation training, the above cases prove that brain visualization is not only a tool to "display data" but also a bridge connecting scientific discovery, clinical practice, and public perception. Translating abstract neural activity into perceptible images through visual language, Interactional models, and traceable, dynamic processes allows researchers to formulate hypotheses more efficiently, doctors to formulate solutions more precisely, learners to understand how thoughts are produced more intuitively, and ultimately drive brain science from the "lab" to the "real world."

Chapter 9

Geoscience and Environmental Visualization

9.1. Introduction: When Natural Mysteries Meet Digital Visualization — Decoding the "Visual Key" of the Earth's Pulse and Environmental Change

Earth, the blue planet that hosts life, is a complex and elaborate giant system — the atmosphere, the hydrosphere, the lithosphere, and the biosphere are interwoven and maintain a dynamic balance through the circulation of material (e.g., carbon cycle and water cycle) and energy flow (e.g., solar radiation-driven climatic processes). From soil particulate adsorption of pollutants at the microscale to global climate change at the macroscale, from transient seismic wave transmission and volcanic eruptions to long-term glacial melting and sea level rise, from air pollution and deforestation caused by human activities to plate tectonic plates and ocean circulation led by natural processes, etc., much of this critical information about the health of the planet and the sustainability of humankind is in the form of massive, multi-dimensional, and dynamic data: Weather satellites send back global cloud maps and temperature data every minute. Seismological stations record small vibrations in real time of the movements of the Earth's crust, marine buoys continuously monitor the temperature, salinity, and flow rate of seawater, and environmental sensor networks capture subtle changes such as PM2.5 concentrations in the air and soil moisture (Zhang *et al.*, 2021).

The core mission of Earth science and environmental visualization is to combine these "invisible" Earth system processes with "hard to understand" environmental data. Translated into intuitive and perceptible visual expression, it has become an important tool for scientists exploring the laws of the Earth, policymakers formulating environmental policies, and the public understanding sustainable development. It acts as a digital eye for the Earth, through multiscale integration (from local to global), dynamic representations (such as the temporal evolution of weather processes, long-term changes in geological formations), and multivariate correlations (such as the coupling of climatic factors and ecological responses). Let the Earth's "breath," "pulse," and "scar" be clearly visible — scientists can visually track the spatial and temporal distribution of atmospheric carbon dioxide concentration, revealing the driving force of global warming. Engineers can visualize changes in the path and intensity of typhoons and deploy disaster prevention and mitigation measures in advance. The public can participate in low-carbon lifestyle choices by visualizing the air quality index around them.

In the field of earth sciences and the environment, the value of visualization technologies cuts across the whole chain of scientific research, disaster warning, resource management, and public education:

- In scientific research, it is a "microscope" and a "telescope" that reveal the laws of the Earth. For example, by visualizing the chemical diffusion processes of deep-sea hydrothermal vents, marine scientists can explore the origins of life in extreme environments. By simulating the stress accumulation and release of plate motion in three dimensions, geologists are able to predict potential risk areas for earthquakes. By integrating satellite remote sensing with ground-based observations, climatologists can analyze the correlation between changes in global vegetation cover and the capacity of carbon sinks. These visualizations help researchers to not only overcome the limitations of traditional observation methods (such as the difficulty of deep-sea exploration and the processing of long-term sequences of data) but also discover hidden patterns of the Earth system through the fusion of multidisciplinary data (such as linking meteorological and ecological data) (Liu *et al.*, 2020).
- It is the "early warning radar" that guards lives and property in the event of disasters. Natural disasters such as earthquakes, floods,

typhoons, and landslides are sudden and destructive, and timely and accurate early warning is the key to reducing losses. The visualization technology enables seismic station networks to monitor seismic source parameters (e.g., magnitude and central location). Weather satellites capture typhoon paths and wind speeds, and geological radars scan landslide hazard points, and other information in the form of dynamic maps, timeline animations, and risk thermographs, to help decision makers quickly assess the scope and urgency of disasters and formulate scientific evacuation and rescue plans. For example, by visualizing the inundation extent and flow rate of a flood in real time, emergency management can pinpoint residential areas that need to be evacuated. By simulating the distribution of wind and rain after a typhoon makes landfall, city planners can optimize the layout of wind protection facilities (Kumar *et al.*, 2022).

- In resource management, it is a "policy advisor" for optimal utilization and conservation. The Earth's resources (e.g., water resources, mineral resources, and forest resources) are finite, and how to achieve sustainable development and utilization is a key issue. Visualization technology can integrate multiple source data (such as land use change from satellite remote sensing, ground monitoring of water resources reserves, and statistical data mining), through dynamic charts to show the temporal and spatial distribution of resources and consumption trends and assist managers in developing scientific allocation and protection strategies. For example, by visualizing the distribution and demand of global water resources, water shortage hotspots can be identified, and interbasin water transfer projects can be planned. By simulating the impact of deforestation on biodiversity, it is possible to optimize the areas and intensity of forestry exploitation and balance economic development with ecological protection. In public education, it is a "bridge of knowledge" to promote environmental awareness. Earth sciences and environmental issues (such as climate change, plastic pollution, and biodiversity loss) are relevant to everyone's lives, but complex scientific concepts often feel alien to the public. Visualization techniques can turn these abstract questions into vivid visual stories — animations showing how melting glaciers are causing sea levels to rise, interactive graphs showing the relationship between individual carbon emissions and global warming, and VR experiences simulating breathing on smoggy days. These intuitive expressions not only lower the threshold of public understanding but also inspire people's sense of concern and responsibility for the Earth's environment, driving the transition from "cognition" to "action (Gupta *et al.*, 2021)."

This chapter focuses on typical scenarios and core technologies for geoscience and environmental visualization, showing, through a rich case study, how visualization technologies are driving progress in this field at different levels: from dynamic monitoring of global climate change to intelligent early warning of geological disasters, from the health assessment of marine ecosystems to mitigation strategies for urban heat island effects, and from regular exploration by professional researchers to the improvement of environmental awareness by the public. These practices not only reflect the scientific value and social significance of visualization technology but also reveal a central law — when digital visualization meets the Earth system and when complex environmental data are combined with intuitive expression, we have a "digital key" to understanding the mysteries of our planet and protecting our homeland (Hansen *et al.*, 2023).

9.2. Earth Insights into the "Digital Eye" of the Earth System

The core objective of earth science and environmental science is to understand the Earth's various spheres (atmosphere). The mechanics of the hydrosphere, the lithosphere, and the biosphere are studied, monitoring environmental processes and predicting future trends, ranging from microscopic soil particles and atmospheric molecules, to macroscopic global climate systems and plate tectonic movements, to geologic and ecological evolution at long scales. These objects are remarkably complex — Earth system layers exchange material (such as the carbon cycle). The water cycle is closely linked to energy transfer (e.g., solar radiation and geothermal activity), and environmental processes involve both rapid dynamic events (e.g., typhoons, floods, and earthquakes) and slow long-term evolution (e.g., glacier retreat, sea level rise, and soil erosion). Traditional research relies on field observation (e.g., weather stations and seismic stations), experimental simulation (e.g., climate patterns and geological experiments), and sampling analysis (e.g., soil composition and water quality testing) but faces a large number of multiple sources, and with multi-scale Earth science data (such as satellite remote sensing image, numerical simulation output, and sensor network data), researchers and policymakers need a tool that can "break through the barriers of time and space and integrate multi-dimensional information," which is the core value of earth science and environmental visualization (Zhao *et al.*, 2023).

The essence of geoscience and environmental visualization is to abstract data in the Earth system (e.g., temperature, atmospheric pressure, soil moisture, geological stress) are translated into intuitive visual expressions (such as graphs, images, dynamic simulations, or interactive interfaces), through multi-scale integration (from local to global) and dynamic presentation (such as the time-series evolution of weather processes). Long-term changes in geological formations are linked to multiple variables (such as the coupling of climate factors and ecological responses) to help researchers shed light on how the Earth works, support environmental policymakers in developing scientific strategies, and communicate sustainable development ideas to the public. Its application scenarios run through basic research (e.g., climate change mechanism analysis), applied practice (e.g., disaster warning, resource exploration), policy formulation (e.g., carbon neutrality target assessment), and public education (e.g., environmental awareness) (Wang *et al.*, 2022).

This section focuses on the application scenarios of earth science and environmental visualization. Three core content areas are supported by digital media technology and typical practice cases, and the system analyzes how it can become a "digital eye" for insights into the Earth system through dynamic display of space-time, multi-layer correlation analysis, and immersive experience.

9.2.1. *Application Scenarios for Earth Science and Environmental Visualization: Multi-Scale Requirements from Micro-Processes to Global Systems*

Earth science and environmental research cover multiple scales and processes, and the requirements for visualization in different scenarios have commonalities (e.g., accuracy and timeliness) and different focuses due to differences in subjects and objectives. The following are typical application scenarios and their core requirements:

1. Atmospheric and Climate Visualization: Monitoring Weather and Predicting Climate

Scenario: The meteorological department monitors the state of the atmosphere (e.g., temperature, humidity, and wind speed) in real time and predicts short-term weather (e.g., heavy rain). Climate scientists analyze long-term climate trends (such as global warming and changes in the frequency of extreme events) and assess the impact of human activities (such as greenhouse gas emissions in relation to warming).

Core requirements: These demonstrate the temporal and spatial distribution of atmospheric variables (e.g., dynamic changes in global temperature fields), dynamic processes of weather systems (e.g., the formation and movement of cyclones), long-term trends in climate factors (e.g., the rate of sea surface temperature increase over nearly a century), and linkages between human activities and climate (e.g., the superposition of carbon emission hotspots and warming regions).

2. Hydrology and Ocean Visualization: Tracking the Flow of Water and Understanding the Ocean

Scenarios: Water engineers monitor river levels, flows, and water quality (e.g., flood warning and water resource management), and marine scientists study ocean circulation (e.g., El Niño), seabed topography (e.g., trench and ridge distribution), and marine ecology (e.g., coral reef degradation).

Core requirements: To present spatial distribution and dynamic changes in water bodies (e.g., real-time water levels of rivers and surface flow rates of oceans), and the temporal evolution of hydrological processes (e.g., the ebb and flow of a flood and the recharge and excretion of groundwater), the vertical structure of the marine environment (e.g., changes in seawater temperature with depth), and the interference of human activities with hydrological systems (e.g., the impact of dam construction on river ecology).

3. Geology and Disaster Visualization: Structure Analysis and Risk Prevention

Scenarios: Geologists study the structure of the lithosphere (e.g., plate boundaries and fault distribution), the distribution of mineral resources (e.g., deposits of oil and metals), and disaster managers monitor geological hazards such as earthquakes, landslides, and mudslides (e.g., disaster warning and emergency response).

Core requirements: These demonstrate the spatial distribution of geological formations (e.g., the type of plate boundaries and activity intensity), vertical stratification of subterranean structures (e.g., the crust, mantle intersections and rock types), the location and level of risk of the potential hazard point (such as the slope of the landslide prone area superimposed with rainfall), and the dynamic modeling of the hazard course (such as the seismic route and the debris flow range).

4. Ecological and environmental visualization: Assessing state and guiding conservation

Scenarios: Ecologists monitor biodiversity (e.g., species distribution and habitat fragmentation). Policymakers assess the state of health of ecosystems (e.g., forest cover change and wetland degradation) and develop conservation strategies (e.g., natural conservation area planning).

Core requirements: To present the spatial pattern of ecological systems (such as the distribution of different vegetation types), spatiotemporal changes of environmental factors (such as regional differences and seasonal fluctuations of PM2.5 concentrations), ecological linkages between human activities (e.g., the encroachment of agricultural land by urban expansion), and assessment of the effectiveness of conservation measures (e.g., the increase in carbon sinks after afforestation).

9.2.2. *Core Support for Digital Media Technologies: From Data Acquisition to Immersive Experiences*

The efficient implementation of geoscience and environmental visualization relies on several key capabilities of digital media technologies, which provide the underlying support for multi-source data integration, dynamic process simulation, and cross-scale interaction:

1. Multi-source data fusion and processing technique

- **Remote sensing and sensor data:** These include satellite remote sensing images (e.g., global LST, vegetation index data from MODIS, and Landsat), a network of ground sensors (such as hourly temperature/

humidity records from weather stations and water temperature/salt monitoring from ocean buoys), and drone aerial data (such as high-resolution topographic images). It needs to be processed by geometric correction (eliminating sensor angle deviation), radiological scaling (uniforming data matrix), and splice fusion (multiple images synthesizing complete regions) to form a continuous spatial–temporal dataset.

◆ **Numerical simulation outputs:** These include high-resolution grid data generated by climate models (such as the CMIP6 global climate model), hydrological models (such as the SWAT watershed hydrology model), and geological simulations (such as a numerical simulation of plate movements) (such as global 1 × 1 km temperature lattice). The soil moisture value per 100 × 100 m in the watershed needs to be optimized by subtracting (extraction of key areas), inserting (filling missing grids), and formatting (adapting visualization tools) to ensure that the data can be processed efficiently (Eyring *et al.*, 2021).

2. 3D Earth Modeling and Visualization Technology

◆ **Digital Earth Platform:** Based on geographic information system (GIS) and 3D earth engine (such as Google Earth Engine and Cedium JS), human elements (such as cities and transport networks) and scientific data (such as temperature, vegetation cover) are superimposed to create interactive "digital Earth" models that allow users to zoom from a global perspective to a local area (such as focusing from the global sea temperature distribution to the El Niño phenomenon in a certain sea area).

◆ **3D modelling of geological bodies:** Through geologic borehole data (such as rock types at different depths), seismic reflection profiles (such as structural stratification within the Earth's crust), and drone tilt photography (such as surface morphology in mines), three-dimensional modeling software (such as ArcGIS Pro and 3DMine) is used to build a three-dimensional model of underground geological structure (such as the spatial distribution of faults and veins) to assist resource exploration and disaster assessment (Cesium, 2022).

3. Dynamic visualization and simulation techniques

◆ **Simulation of weather and climate processes:** Based on hourly output data from Numerical Weather Prediction (NWP) models (e.g., ECMWF and GFS), the evolution of weather systems (e.g., typhoon formation → intensification → landfall, including cloud range, wind speed variation and precipitation distribution) is dynamically demonstrated. Based on long-term simulations of climate models, such as future warming trends in the RCP scenario, timeline sliders show changes in global temperature, sea level curve, and frequency of extreme events (e.g., heatwaves and droughts) from nearly 100 to 100 years into the future.

- **Dynamic modeling of geological hazards:** Finite element analysis (FEA) or discrete element method (DEM) is used to simulate the propagation of seismic waves (e.g., the distribution of epicentral and peripheral intensities due to the differential velocities in different rock formations). The initiation and flow of landslides (e.g., the process of instability after rainfall infiltration leads to saturation of the slopes), the mixing and washing of sediment flows (e.g., the interaction of solid particles with water flows), and dynamic graphics (such as the concentric circle of seismic waves and the brown fluid spread of mudslides) are used to aid disaster warning and contingency planning.

4. Interaction and Immersive Technologies

- **Interactive exploration:** The user can click on a specific area on the map (such as an air quality monitoring point in a city) and view historical data at that point (e.g., changes in PM2.5 annual average), real-time monitoring values (e.g., current AQI index), and associated factors (e.g., distribution of emissions sources at surrounding plants). Through multi-view interactions (such as showing a temperature distribution and a wind field vector at the same time), the dynamics of weather systems (such as facing formation due to the convergence of cold and warm air) are analyzed.
- **Virtual Reality (VR)/Augmented Reality (AR):** In geology teaching, students use VR devices to "enter" the Earth's crust and observe the stratified structure of the lithosphere (such as the boundaries and material composition of the crust, mantle, and core). During a disaster drill, rescuers improved emergency response efficiency by layering AR glasses on the distribution of damage to buildings after the earthquake (red-marked collapsed buildings), rescue routes (green arrows), and dangerous areas (yellow alert strips).

9.2.3. *Typical Practice Cases: Visualization Technology-Driven Geoscience Discovery and Environmental Decision Making*

Case 1: Global Dynamic Climate Change Visualization Platform — a bridge from data to decision-making

In order to address the challenge of global warming, an international scientific research institute has jointly developed a global climate change visualization platform. The goal is to turn complex climate data — such as nearly a century of recorded global temperatures, sea levels, and extreme events — into intuitive visualizations that will help policymakers understand climate change trends, assess the effectiveness of emission reduction policies, and communicate scientific information to the public.

The visualization scheme:

- **Multiple time-scale climate trends display:** Through an interactive timeline (1880–present), users can dynamically view global average annual temperatures (red curve), the trend of sea surface temperature (blue curve), sea surface height (green curve), and the temperature rise rate in different time periods (e.g., preindustrial vs. 21st century). Overlay key event markers (such as the signing of the Kyoto Protocol in 1997 and the entry into force of the Paris Agreement in 2015) to analyze the potential impact of policy interventions on climate change.
- **Analysis of spatial distribution and regional differences:** Using a global map as a background, a color gradient (blue → red) shows the spatial distribution of global temperature increases over the past 30 years (e.g., 3–4°C in the Arctic, much higher than the global average of 1.1°C), and by clicking on the area to pop up detailed data (such as "a station in the Arctic Circle: annual average temperature increase of 2.8°C from 1990 to 2020, more significant warming in winter"), the combination of population density (grey shade in densely populated areas) and carbon emission intensity (black marking in industrial areas) reveals a "high correlation between concentrations of human activity and hot spots of warming."
- **Dynamic simulation of extreme events:** These include dynamic playback of historical cases for extreme events such as typhoons,

heatwaves, and droughts (such as the path evolution of Hurricane Katrina in 2005, e.g., the frequency of extreme heat events in the world by 2100 under the RCP8.5 scenario is three times higher) to assist decision makers in disaster preparedness planning (e.g., enhancing coastal flood control facility and optimizing urban heat island mitigation measures).

- **Public interaction and science modules:** These include setting up a "carbon footprint calculator" (where users input daily behaviors such as "how many flights a year" and "how much electricity they use for their homes"), and the system calculates individual carbon emissions and recommends actions to reduce them). The "Future Scenario Experience" (in which users select a "high reduction" or "low reduction" scenario to observe differences in global sea level and food production in 2050) promotes public responsibility for climate change through gamification and immersive design.

Digital media technology applications: The platform integrates satellite remote sensing data (such as NASA's Global Temperature Products), climate model output (such as the CMIP6 multi-model collection), and ground-based observation networks (such as a 100-year record from the Global Weather Station) and allows smooth global-scale rendering through WebGL technology. These support real-time interactions of tens of millions of data points (e.g., users zoom down to a city level to still see detailed weather station data) and provide "immersive climate tours" through VR models (e.g., "traveling to Shanghai in 2100 to experience flooded streets after sea level rise").

Case 2: Intelligent Early Warning Systems for Geological Disasters — Visual Support from Risk Identification to Emergency Response

The scene: The mountainous region of southwest China (such as Sichuan and Yunnan) is characterized by active geological formations and frequent earthquakes, landslides, mudslides, and other disasters. A provincial natural resources department has developed an "intelligent early warning system for geological hazards" with the goal of identifying high-risk areas in advance and guiding emergency evacuation through real-time monitoring data (such as rainfall, land stress, and slope displacement) and visual analysis of historical disaster data.

The visualization scheme:

- **Multi-source data for real-time monitoring:** This includes integrating real-time rainfall (hourly updates), GNSS monitoring equipment's slope displacement data (millimeter-level accuracy), seismic station records of microearthquake activity (such as the frequency of small earthquakes and the depth of the earthquake source), and showing trends in key parameters through dynamic dashboards (such as columnar growth of rainfall in real time, and displacement curves showing tiny changes in the slope), superimposed geological map (marked fault and rock type), land use map (distinguish forest land, farmland, and residential area), and clear the potential impact of disasters.
- **Dynamic assessment of level of risk:** The level of risk (low/medium/hithe/very high) of each slope or area was calculated, and the level of risk (green → yellow → orange → red) was marked directly. High-risk area (red) automatically triggers an early warning pop-up window, showing that "it is expected that the rainfall in the next 24 hours will exceed 100 mm, and the probability of landslide in this area is >70%, and it is recommended to evacuate the surrounding residents immediately."
- **Disaster Process Simulation and Contingency Planning:** Discrete Element Simulation (DEM) is used to demonstrate the process of instability (such as crack propagation). Blocks are sliding, and dynamic graphics (brown earth blocks gradually declining) are used to aid in the analysis of disaster impact areas (e.g., roads that may bury villages downstream); combine digital elevation model (DEM) data to plan optimal evacuation routes (green arrows represent safe paths, avoiding steep slopes and rivers), and overlay real-time road conditions to rescuers via AR devices (such as red markings of sections blocked by mudslides).
- **Retrospective and learning from historical cases:** These include nearly 10 years of typical disaster cases in the region (e.g., landslides triggered by the Lushan earthquake in 2013) and show a timeline playback of anomalies in monitoring data prior to the disaster (e.g., sudden increases in rainfall and accelerated displacement rates). Dynamic processes at the time of a disaster (e.g., the initiation and movement of landslides) and its aftermath (e.g., damage to homes and road disruptions) help researchers to formulate patterns (e.g., "landslide risk increases significantly when rainfall exceeds 200 mm for three days in a row") and optimize early warning thresholds.

Digital media technology applications: The system integrates IoT sensor networks (transmit monitoring data in real time), GIS spatial analysis (calculate buffers and impact areas of risk areas), and 3D visualization engines (show the 3D shape of the slopes and the path of disaster flows). Through the mobile APP (to support front-line staff to view real-time risk during field inspections) and the large screen of the Zhizhicenter (multi-view linkage display of global risk and emergency resource distribution), the entire process of monitoring–assessment–early warning–response was supported by visual support, which significantly reduced the number of human casualties and property damage from geological disasters.

Case 3: Dynamic monitoring and conservation planning of marine ecosystems — from data insights to management decisions

The oceans cover about 71% of the Earth's surface and are the largest ecosystem on the planet but face multiple threats from overfishing, marine pollution (e.g., plastic waste and oil spills), and climate change (e.g., acidification and warming). An international marine conservation and protection organization has developed a "marine ecosystem monitoring and protection planning system." By integrating multiple ocean data (such as satellite-derived SST and chlorophyll, buoy-derived DO and pH, and UAV-derived RR), the state of health of the ocean ecosystem can be visualized to help formulate scientific conservation strategies.

The visualization scheme:

- **Multivariate marine environment monitoring:** Using a global ocean map as a background, a color gradient shows the temporal and spatial distribution of key environmental parameters — sea surface temperature (blue → red indicates low to high temperature reveals marine heatwave zones), chlorophyll concentrations (green to yellow indicates low to high, reflecting phytoplankton biomass), dissolved oxygen levels (purple to blue indicates low and high, associated with fish habitats), and pH values (red to blue indicating acidification → normal, identifying water acidification hotspots). For example, the sea off the Equator exhibited significant sea surface temperature increases during El Niño years (red area expansion) along with unusually low chlorophyll concentrations (green sparseness), illustrating the impact of climate change on primary productivity.

- **Marine life distribution and habitat visualization:** Tracking (e.g., whales), the migration paths of sea turtles, underwater drone capture (e.g., spatial distribution of coral reefs and seagrass beds) and acoustic monitoring (e.g., cluster activity of fish) overlay the range of marine life and key habitats (e.g., spawning areas and feeding areas) onto environmental parameters. For example, coral bleaching (whereby symbiotic algae detach due to high sea temperatures) is marked on coral reef distribution maps with white plaques and associated with anomalies in sea surface temperature in corresponding regions (red high-value regions), revealing a direct correlation between temperature increases and biological degradation.
- **Conservation planning and effectiveness assessment:** This is based on ecological connectivity analysis (such as species migration pathways between habitats) and conservation prioritization models (such as integrating biodiversity, threat levels, and conservation costs). The optimal layout of marine conservation and protection areas (e.g., the proposed urban conservation area with green polygons) is recommended. The effectiveness of conservation measures is assessed by dynamic comparisons. Users can adjust the scope and boundaries of the urban conservation area to see the impact on the connectivity of the habitat of key species in real time (such as "expanding the conservation area will increase the connectivity of whale migration paths by 40%").
- **Public participation and education modules:** A visual dashboard of the Ocean Health Index (combining environmental parameters and biological indicators to indicate the state of the region's marine ecosystems on a scale of 0–100 points) is set up, allowing the public to view the health scores of their nearby seas by clicking on a map and raise awareness of marine conservation and protection through VR experiences such as "Dive Into Coral Reef Ecosystems" (observing healthy and degraded corals) or "Virtual Ocean Cleanup" (simulating plastic waste removal and viewing the impact on marine life).

Application of digital media technology: The system integrates multi-source satellite remote sensing data (such as MODIS and Sentinel-3), buoy and UAV real-time monitoring network, as well as historical ecological survey data (such as nearly 20 years of coral reef census records). 3D earth engines (such as CesiumJS) enable interactive browsing of the world's oceans, allowing users to zoom in to a local area (such as a coral cluster) to view images and ecological parameters in centimeter-resolution and use dynamic timelines (showing trends in marine environment and biological indicators over the past 10 years) and multi-view interfaces (synchronized display of environmental parameter charts, biological distribution maps, and conservation planning diagrams) provide marine scientists, policymakers and the public with a "one-stop" tool for marine ecological management.

Case 4: Precision monitoring and mitigation strategies for urban heat island effects — from micro-perception to macro-regulation

Scenario: With the acceleration of urbanization, the urban heat island effect (the phenomenon of urban centers with significantly higher temperatures than the surrounding suburbs) intensifies, leading to frequent summer heat disasters, increased energy consumption (such as increased air conditioning use), and increased health risks (such as heat exposure, cardiovascular disease). The meteorological and planning departments of a megalopolis jointly developed the "Urban Heat Island Effect Fine Monitoring and Mitigation Strategy Visualization System," which visually analyzes the spatial distribution, driving mechanism, and mitigation effect of urban heat island through high-density sensor networks and multi-source data fusion.

The visualization scheme:

- **Temporal and spatial dynamics of heat island intensity:** Based on micrometeorological stations (1–2 per square kilometer) distributed throughout the city and thermal infrared data (resolution up to 10 m) from satellite remote sensing, this monitors the surface temperature of a city in real time (expressed in a color gradient, with red to blue indicating high to low temperature), and demonstrates the spatial–temporal evolution of heat island intensity through dynamic timelines (such as changes from day to hour in summer) — e.g., during the working day, the surface temperature of urban centers (such as commercial areas and transportation hubs) is 5–8°C higher than that of the suburbs, and the high temperature areas are highly consistent with the distribution of traffic flow and population density. The nighttime heat island effect is even more pronounced (cities cool down more slowly than suburbs), creating a persistent "hot core."
- **Drive factor analysis:** This correlates land surface temperature with multiple source data (e.g., land use type (green, concrete, water),

building density, population density, and traffic flow) and performs spatial superposition and statistical analysis to reveal key drivers through heat and dispersion maps — e.g., heat island strength in regions with a building density of >60% is on average 3°C higher than in low-density areas, while local temperatures can be reduced by 1–1.5°C for each 10% increase in greenfield coverage. Around major arterial roads with high traffic (e.g., with more than 5,000 cars per hour), the temperature is significantly higher than on adjacent streets.

- **Mitigation simulation and evaluation:** Multiple mitigation options are built into the system (such as increasing rooftop greening, expanding city parks, optimizing road materials (such as the use of high-reflective materials)), comparing numerical simulations (such as surface temperature predictions based on energy balance models) with visualizations (such as "Option A: After the green area on the roof is increased by 20%, A 1.2 C drop in core temperature." "Option B: A new 5-hectare park can reduce temperature within 300 meters of the perimeter by 0.8 C"), which assists decision makers in selecting the optimal strategy combination. Users can adjust the parameters (such as greenery coverage ratio and park location) through the interface, real-time view of the improvement effect on the intensity of the heat island (such as "after a concrete plaza in a commercial area to permeable green space, the summer heat peak temperature in the area decreased by 2°C").
- **Public Services and Health Warning:** These provide citizens with real-time heat island maps (check the temperature of the surrounding area through the mobile APP) and push personalized advice when high temperature alerts ("Your area is currently 4°C warmer than the surrounding area, so it's advisable to avoid going out at noon and head to a nearby park for the summer"), and use health data (e.g., number of heat stroke emergency cases) to analyze the relationship between heat islands and public health (e.g., "5% increase in the incidence of a disease for every 1°C increase") to promote "climate-friendly" urban planning.

Digital media technology applications: The system integrates a high-density sensor network (real-time transmission of temperature, humidity, and wind speed data), satellite remote sensing (to provide a large range of thermal infrared images), and geographic information system (GIS) spatial analysis, through three-dimensional city model (superimposed building, green space, roads, and other elements) intuitive display of the spatial structure of the heat island and uses dynamic visualizations (such as real-time temperature fluctuation curves and temperature comparison animations before and after mitigation) and interactive queries (where users click on any block to see detailed environmental parameters and mitigation potential). It provides city managers, planners, and residents with a comprehensive "sense-analysis-decision-action" tool to help cities address the challenges of climate change.

Common Insights of Geoscience and Environmental Visualization Cases

From global climate change to the urban heat island effect, the four new cases (marine ecosystems and urban heat islands) and the original cases (climate change platforms and geological disaster warning) jointly demonstrate the core value of visualization technology:

- **Multi-scale integration:** This shows both global-scale macro trends (such as rising global temperatures and ocean acidification) and micro-level details of local areas (such as the bleaching plaques of a coral reef and temperature anomalies in urban neighborhoods).
- **Dynamic process presentation:** This reveals the dynamic properties of the Earth system through timeline animation (e.g., time series of climate change, seasonal fluctuations in marine ecosystems) and real-time monitoring (e.g., real-time updates of rainfall from geological disasters, hourly temperature changes in urban heat islands).
- **Multivariable association:** This is the analysis of associations between environmental variables (e.g., temperature, pH, and chlorophyll concentration) and human activities (e.g., carbon emissions, land use, and urban planning) or biological responses (e.g., species distribution and coral health) to uncover hidden cause and effect relationships.
- **Decision Support and Public Participation:** This is to provide policymakers with scientific evidence (e.g., urban conservation area optimization and mitigation options) and to increase public awareness and participation through interactive tools (e.g., carbon footprint calculator and Heat Island Maps app).

Core values of geoscience and environmental visualization
The essence of earth science and environmental visualization is to transform "invisible" (such as subsurface geological structure and general atmospheric circulation dynamics) into "visible" and "complex" (such as multi-layer correlation and long time-scale evolution) into "understandable" through digital media technology. It is not only a "digital microscope" and "space-time telescope" for researchers to explore the mysteries of the Earth but also a "scientific navigator" for policymakers to formulate sustainable development strategies. It is also a "visual classroom" for the public to communicate the concept of ecological progress. With the development of artificial intelligence (such as deep learning to predict extreme weather), digital twins (such as building virtual Earth systems for real-time synchronized reality), and immersive technologies (such as VR to simulate geological disaster escape), the future of earth science and environmental visualization will move toward greater accuracy (e.g., kilometer-scale real-time monitoring). Deeper layers (such as the coupling mechanisms of Earth system layers) and broader participation (such as real-time public interaction and feedback) are evolving to provide more powerful cognitive tools and decision support to safeguard humanity's common home planet. In the following chapters, we further explore the cross-disciplinary appeal of scientific visualization in other typical scientific fields, such as materials science and energy science.

9.3. Environmental Visualization: Decoding the Ecological Picture of the Anthropocene Crisis

In the field of earth scientific visualization, "environmental visualization" focuses on the interaction between human activities and the natural environment and reveals the dynamic changes of environmental elements such as air, water, and soil, as well as the response mechanism of ecological system to stress through visual language. In contrast to traditional geosciences, which focus on the macroscopic aspects of "geo-physical-climate" systems, environmental visualization places greater emphasis on microscopic perceptions of "human scales" (e.g., community-level pollution distribution) and individual exposure risks (e.g., climate change–biodiversity–human health chain feedbacks) presented with complexities of "cross-system coupling." This section demonstrates, through technical analysis and typical examples, how visual design can transform "invisible environmental crises" into "perceptible, traceable, actionable" scientific images.

9.3.1. *Core Challenges and Objectives of Environmental Visualization*

The nature of environmental problems is multi-media coupling (transport and conversion of pollutants in air, water, and soil) and temporal and spatial heterogeneity (e.g., the concentration of PM2.5 varies significantly between urban heat island areas and suburbs). The fact that heavy metal pollution is patchy around industrial sites and that it is both long-term cumulative and sudden (e.g., ozone depletion is a decades-long cumulative effect, while chemical spills are transient exposures to high concentrations) make their visualization a challenge that must be addressed simultaneously:

- **Data fragmentation:** Environmental monitoring data come from a variety of sources (satellite remote sensing, ground sensors, and mobile monitoring equipment) and are not uniformly formatted or accurate (e.g., satellite data with low spatial resolution but wide coverage, portable sensor data with high accuracy but small range).
- **Hidden causality:** The relationship between environmental change and human activities is often covered up by intermediate variables (such as deforestation → water loss and soil erosion → the increase of river silt → the destruction of aquatic habitat).

The public perception gap: The public lacks an intuitive understanding of professional indicators such as "ppm (parts per million)" and "AQI (air quality index)," and it is difficult to relate abstract data to their own health.

Its core objectives are achieved through visual design:

- **Process transparency:** Tracking the migration path of pollutants from source (e.g., plant discharges and fertilizer application on agricultural land) to receptors (e.g., humans and ecosystems).
- The "superscale" and "density value" were transformed into "incidence of a disease" and "number of dead fish."

- **Decision support:** This provides policymakers with "comparisons of environmental improvements under different emission reduction scenarios" and the public with "advice on actions to avoid high-risk areas."

9.3.2. *Key Technical Paths and Typical Scenarios for Environmental Visualization*

Scenario 1: Spatiotemporal dynamic tracking of atmospheric pollution — from "smog map" to "exposure risk profile"
Objective: To reveal the spatial and temporal distribution of pollutants such as particulate matter (PM2.5/PM10), ozone (O_3), and nitrogen dioxide (NO_2) in air, as well as the differentiated characteristics of exposure risk in populations.

Technology implementation:

- **Multi-source data fusion:** This is the integration of satellite remote sensing data (e.g., MODIS for aerosol optical thickness monitoring), spatial resolution of 1 km, data from ground-controlled national control sites (updated hourly with street-level accuracy), and mobile sensor networks (such as portable monitors on board vehicles/drones that capture microscale changes).
- **Dynamic visual design:**
  - **Thermodynamics maps of space-time:** This shows the spatial distribution of pollutant concentrations in color gradients (e.g., red = heavy pollution and yellow = light pollution) (e.g., the "flake high zone" of winter PM2.5 in Beijing–Tianjin–Hebei). Historical changes can be traced back through the time axis slider (such as "Beijing PM2.5 average annual concentration in the past 10 years from 89 $\mu g/m^3$ to 38 $\mu g/m^3$").
  - **Exposure risk profile:** This overlays population density data (e.g., elderly population and distribution of children) with activity trajectory (e.g., commuter routes, periphery of schools) and labels "high risk areas" (e.g., kindergartens next to major morning and evening rush roads and NO_2 concentration more than twice the national standard).
  - **Source-tracing simulation:** The main sources of pollutants are animated through reverse proliferation models (such as the CMAQ

chemical transport model) ("local motor vehicle emissions contribute 40%, surrounding province transport contributes 30%").

Typical case: In the "Beijing air quality" app launched by the Beijing Municipal Bureau of ecological environment, users can view the location of the PM2.5 density (number + color bar), for the next 3 hours (up/down arrow), by click "Source Analysis of Major Pollutants" (e.g., "Secondary minerals account for 55% of PM2.5, suggesting area of interest in transport and nitrates formation").

Scenario 2: Watershed-level traceability and treatment tracking of water pollution — from "black and foul water" to "eco-remediation effectiveness"

Objective: To analyze the sources of pollutants (e.g., heavy metals, organic pollutants, and nutrient salts) in surface water (rivers and lakes) and groundwater, to track their diffusion along water flow paths, and to visualize the effects of treatment projects (e.g., wastewater treatment plant upgrades and wetland construction).

Technology implementation:

- **Data acquisition:** Multi-dimensional data are obtained from water quality monitoring buoys (real-time monitoring of pH, dissolved oxygen, ammonia, and other indicators), unmanned ship onboard sensors (collecting samples of river bottom sediments), and groundwater monitoring wells (long-term tracking of nitrate concentration changes).
- **Visualization strategy:**
  - **Pollution map of river basin:** This labels the pollutant density of each section with different colors (e.g., "TP > 0.3 mg/L is the high risk area of eutrophication") and labels the main source of pollution (e.g., aquaculture farms on the upper reaches of the river and printing and dyeing plants along the river).
  - **Migration path simulation:** This uses hydrodynamic models (such as MIKESHE) animated to show how pollutants spread with water flow (e.g., "Benzide leaked from a chemical plant spread from the upstream of the river to a water catchment 15 km downstream in 3 days").
  - **Pre- and post-treatment comparison:** Images and key indicator changes (e.g., "ammonia concentration decreased from 12 mg/L to 1.5 mg/L") are shown side by side with images and changes in key

indicators before treatment (e.g., black and smelly water bodies with dissolved oxygen levels below 2 mg/L, leading to fish mortality).

Typical example: A project to visualize the pollution control of the "mother river" in a city of the Yangtze River through a drone aerial comparison (waste floating in the river in 2015 and ecological floating islands in 2022). The timeline of water quality indicators (COD decreased from 200 mg/L to 30 mg/L) and public participation data (e.g., "Coastal residents' complaints decreased by 72%") intuitively demonstrate the three-step governance logic and effectiveness of "source control, interdiction, internal source governance, and ecological restoration."

Scenario 3: The hidden risks of soil contamination are revealed — from "industrial sites" to "agricultural safety"

Objective: To track the accumulation of heavy metals (e.g., cadmium, lead, and arsenic) and persistent organic pollutants (e.g., polycyclic aromatic hydrocarbons) in soil and to analyze potential threats to crop growth and human health (e.g., cadmium rice incident).

Technology implementation:

- **High-Resolution Sampling:** This is the data on contamination profiles obtained by gridded soil drilling (every 100 × 100 m, to a depth of 0–20 cm), combined with historical land use maps (e.g., the boundaries of lead-zinc mining areas in the 1980s).
- **Visual design:**
  - **Three-dimensional soil cross-section diagram:** This shows the concentration variation of pollutants at different depths (0–10 cm surface layer and 10–20 cm subsurface layer) (e.g., "Cadmium concentration is higher in the surface layer due to pesticide residues, and lower in the subsurface layer due to natural dilution").
  - **Risk transmission chain:** This links arrows to "contaminated soil → crop absorption (e.g., rice has a cadmium enrichment factor of 0.5–1.2) → human intake (calculate daily tolerance of PTDI through intake of staple foods)" and marks key risk points (such as "Cadmium content in paddy soil of a village is 0.8 mg/kg (national standard ≤ 0.3 mg/kg), resulting in 21% excess urine cadmium of local residents").

- **Simulation of restoration schemes:** This compares the costs and effects of different restoration techniques (e.g., guest earth replacement and plant restoration — growing centipede herb to absorb arsenic) (e.g., "The replacement of foreign land needs to invest 500 thousand yuan/ha, and the cadmium concentration will be reduced to below the national standard within 3 years; Phytoremediation takes 5 years, costs only 100 thousand yuan/hectare, but the effect is slow").

Typical example: A visualization report on soil pollution in agricultural land around a mine in Hunan, displayed by linking "pollution map + agricultural testing data" (e.g., "rice fields within 1 km of the mine have excessive cadmium content. The output of rice had a cadmium content of 0.28 mg/kg (the national standard ≤ 0.2 mg/k)"), which prompted the local government to designate "zones where the cultivation of edible agricultural products is prohibited" and to induce farmers to switch to non-food crops such as cotton.

Scenario 4: Coordinated visualization of climate change and biodiversity — from "retreat of glaciers" to "extinction risk"
Objective: To reveal how increasing temperatures and changing precipitation patterns in the context of global warming affect ecosystem structure (e.g., the northward shift of forest belts and coral bleaching) and to assess threats to key species (e.g., polar bears and amphibians).

Technology implementation:
- **Multi-scale data integration:** This includes integrating satellite remote sensing data (such as the MODIS monitoring global vegetation index NDVI), weather station records (changes in temperature/precipitation over the past 100 years), and species distribution databases (such as geographic coordinates of endangered species in the IUCN Red List).
- **Visualization strategy:**
  - **Climate-ecological coupling maps:** These use temperature gradient color bands (blue = low, red = high) to superimpose biome distributions (e.g., tropical rainforests, grasslands, and tundra) and animate vegetation shifts over the past 30 years (e.g., "subtropical forests expanded 150 km to higher latitudes").

- ◦ **Assessment of species vulnerability:** Using a model that combines "climate change exposure (e.g., the rate of rise in habitat temperatures) + ecological adaptability (e.g., flexibility in reproductive cycles)." Label high-risk species (e.g., "the Arctic fox, whose hunting range has shrunk due to the disappearance of sea ice, resulting in a 70% decline in population size over 30 years").
- **Scenario simulations:** These include comparing future ecological projections (e.g., Paris Agreement 2C vs. 1.5C temperature rise) with different emission reduction targets (e.g., "In the 1.5°C scenario, the loss of coral cover is controlled at 10%; Loss of 30% in 2°C scenario").

The World Fund for Nature's (WWF) "Living Planet Report" visualization, through "Polar Bear Habitat Shrinking Animation," "Annual Loss of Tropical Forests (4.1 million hectares in 2022) Dynamic Counter," and "Endangered Species Map" (click to view the distribution of specific species and threat factors), "the abstract climate change impact into a tangible life crisis."

9.3.3. *Design Principles for Environmental Visualization and Public Participation*

Design principles:

- **Living Translation:** This includes transforming professional indicators into familiar reference points for the public (e.g., "PM2.5 concentration of 300 $\mu g/m^3 \approx$ harm from smoking one pack of cigarettes," "1 hectare of forest absorbs 15 tons of CO_2 per year").
- **Spatial–temporal penetration:** This includes simultaneously presenting short-term fluctuations (such as today's AQI) and long-term trends (such as a 10 cm sea level rise in the past 20 years), helping users understand the urgency of the "gradual crisis."
- **Action-oriented:** This embeds "what users can do" in the visualization (e.g., "see the best nearby parks for air quality" and "participate in community garbage sorting points").

Crowdsourcing data (e.g., "Source of pollution photos near my home"), interactive maps (where users mark "areas that feel bad odors" to aid official monitoring), and educational games (such as "Mock Mayor:

Balancing Economic Development and Pollution Control") promote a shift from passive receiving of information to active participation in governance.

Environmental visualization — making crises visible and action feasible. From the "exposure risk portrait" of atmospheric pollution to the "watershed management track" of water pollution, and from the "agricultural product safety correlation" of soil pollution to the "biodiversity crisis warning" of climate change, environmental visualization tears the hidden veil of environmental issues through visual language, transforming scientific data into a perceived threat for the public, a basis for policymakers to rely on, and a target for actors to target. It is not only an important part of earth science but also a key communication tool for promoting "harmony between man and nature" — because only by "seeing" the crisis can we truly "guard" the future.

9.3.4. *Typical Practice Cases: Environmental Visualization — Typical Applications from Monitoring to Action*

The core value of environmental visualization lies in translating abstract environmental crises into a "perceptible, traceable, actionable" scientific picture through visual language. Here are four examples of how visualization can advance scientific understanding and solutions to environmental problems in different scenarios, covering four key areas: air pollution, water treatment, soil remediation, and climate change.

Case 1: Beijing "Blue Sky Defense" air quality dynamic monitoring platform (atmospheric pollution control)
Background: From 2013 to 2017, the average annual concentration of PM2.5 in the Beijing–Tianjin–Hebei region exceeded the national standard by more than twice, and the public's intuitive perception of the "smog siege" and health concerns are strong, and accurate and transparent pollution information transmission is urgently needed.

Visual goal: It is real-time tracking of PM2.5 and other pollutants in time and space, and analysis of the source of pollution contribution to assist the public in avoiding peak travel and the government to accurately implement the policy.

Practical content:

- **Multi-source data integration:** This includes access to 35 state-controlled stations (updated every hour), more than 1000 micro-sensors (covering sensitive areas such as communities and schools), and satellite remote sensing aerosol data (monitoring regional transmission), forming a "macro-micro" joint monitoring network.
- **Dynamic interactive interface:** With the development of "Beijing air quality" official app and WeChat Mini programs, users can view real-time AQI (air quality index) and primary pollutants (such as PM2.5/ozone), by clicking on the map to view the values of nearby monitoring points (such as "Chaoyang District International Trade Site PM2.5 = 78μg/m^3, second-class good") and trace the past 24h concentration curve through the time axis.
- **Pollution tracing and health tips:** Model-based calculations "current PM2.5 levels are influenced by local vehicle emissions (35%) and regional transport (Hebei/Tianjin) (40%)" and personalized advice is given to sensitive populations (children/the elderly) (e.g., "Today's ozone concentration is high, advise to reduce outdoor activities in the afternoon").

Results: After the platform was launched, the public's cognitive accuracy of the causes of pollution increased by 60%, and the number of motor vehicle trips decreased by 12% during peak hours. The government has precisely delineated "key control zones" based on visual data, promoted peak production in steel, cement, and other industries, and helped reduce the average annual concentration of PM2.5 in Beijing from 89.5 μg/m^3 in 2013 to 30 μg/m^3 (a record low) in 2022.

Case 2: A full-scale visualization of the treatment of "black stinking water" in a city in the Yangtze River Delta (water pollution remediation)

Background: Some rivers in this city are blackened and smelly (dissolved oxygen < 2 mg/L, ammonia > 15 mg/L) due to direct discharge of industrial wastewater and mixing of domestic sewage. The citizens' complaints are concentrated, and the visual tracking of the treatment process is needed to verify the effectiveness.

Objective of visualization: To dynamically demonstrate the process of water quality improvement, relate the effects of governance projects (such as sewerage and ecological floating islands) and ecological restoration, and increase public trust.

Practical content:

- **Whole basin monitoring network:** Install online water quality monitoring devices (for real-time transmission of pH, dissolved oxygen, and COD data) at key sections of the river course (upstream/midstream/downstream). Combine this with monthly aerial photography by drones (to capture images of water surface color and distribution of floating objects), thereby creating a dual-dimensional monitoring system for both water quality and landscape.
- **Comparative visualization interface:** A "River Governance One Map" platform was developed, showing on the left an image of a black and foul water body (floating garbage on the surface, dark brown color) before the governance (2018) and key indicators (ammonia = 18 mg/L). The right shows synchronized clear waters (dissolved oxygen > 6 mg/L, ammonia = 0.8 mg/L) after treatment (2022) and shows a timeline animation of the phases of "contamination project completion → siltation and dredging → eco-floating island planting → fish return."
- **Public participation module:** Through the app, citizens can upload river photos and label problems (e.g., "Some section still has a foul odor"), and the system automatically correlates data from recent monitoring points and provides feedback on the progress of the treatment (e.g., "The section of XX that you reported has been scheduled for emergency cleaning and is expected to improve within 3 days").

Results: The visual platform increased public satisfaction with governance projects from 52 to 89%. The river's ecology has been significantly restored (the number of fish species has increased from 2 before the governance to 15), making it a demonstration section of an "urban ecological living room."

Case 3: Risk early warning and decision support system for "soil remediation in mine areas" in Hunan (soil pollution control)
Background: The soil cadmium content of agricultural fields around a historical lead and zinc mine has exceeded the national standard (1.2 mg/kg in some areas, far exceeding the national standard of 0.3 mg/kg), resulting in a high risk of cadmium contamination of rice. It is necessary to visually identify high-risk fields and guide their remediation.

Visual goal: To accurately locate the polluted area, quantify the level of risk, simulate the effect and cost of different remediation programs, and assist the government in formulating scientific governance strategies.

Practical content:

- **High-precision pollution mapping:** This is through gridded sampling (sampling topsoil samples every 50 m × 50 m to detect cadmium/lead/arsenic concentrations), combined with geographic information system (GIS), "soil pollution concentration distribution map" (red/yellow/green marked high risk/medium risk/safe area), superimposed on farmland boundaries and crop types (such as rice fields concentrated in the southern low-lying areas).
- **Risk-health association model:** This calculates the probability of excess cadmium in rice in different regions (e.g., "78% probability of excess cadmium in rice in red zone (0.2 mg/kg)"). The health risks before restoration (estimated affected population of 23,000 after 10 years) and after restoration (target affected population < 500) were compared by histogram.

Comparison and Selection of Restoration Schemes: Visual display of three schemes (foreign soil replacement/cost 500 thousand yuan/hectare, plant restoration/cost 120 thousand yuan/hectare, and the effects of chemical bluntification/cost 80,000 yuan per hectare) (e.g., "plant remediation resulted in a reduction of cadmium concentrations to 0.5 mg/kg within 3 years, while soil replacement achieved the target within 1 year but at a high cost") and the implementation cycle support decision makers in weighing efficiency and economy.

Results: The system helped the government to designate "zones where food produce is banned" (15% of the area) and prioritize the promotion of low-cost plant remediation techniques on high-risk farmland. After three years, the affected population was reduced by 82% and the average soil cadmium concentration fell by 41%.

Case 4: Global interactive early warning platform on climate change and biodiversity (ecological crisis response)

Background: Global warming is already putting one in eight species at risk of extinction, according to a report by the Intergovernmental Panel on

Climate Change (IPCC). However, the public's understanding of how climate change will specifically affect the ecology of the world around it is unclear.

Visual goal: To integrate multi-scale data (global temperature change, species distribution and migration, and urban conservation area effectiveness) to dynamically demonstrate the impacts of climate change on ecosystems and to drive conservation action.

Practical content:

- **Multilevel data integration:** This includes access to NASA global temperature data (1880–2023), the IUCN Red List of Species (containing the geographical coordinates of 38,000 endangered organisms), and satellite-monitored forest/coral reef cover changes (such as 50% loss of global coral reefs over the past 30 years).
- **Interactive scenario simulation:** Users can select a "2°C vs. 1.5°C temperature increase target" to see the corresponding future projections (e.g., "in the 1.5°C scenario, Arctic summer sea ice decreases by 13% every 10 years; in the 2°C scenario, it decreases by 28%") or click on specific areas (such as the Amazon.com rainforest) to observe the feedback cycle of "rising temperature + worsening drought → increasing tree mortality → decreasing carbon sink capacity."
- **Endangered Species Story Library:** This maps habitat changes for high-risk species such as polar bears and Sumatran tigers (e.g., "The polar bear's summer range has shrunk by 30% in the last 20 years") and embeds short videos, such as the "coral bleaching process" captured by scientists, to enhance emotional resonance.

Achievements: The platform was listed by UNESCO as a model tool for climate education. It has received over 5 million global visits. It has encouraged many countries to incorporate biodiversity conservation into their National Intended Contributions (NDC) goals. Public support for reducing carbon footprint increased by 37%.

The power of environmental visualization is from data to action. Together, these cases demonstrate that environmental visualization is not only a "presentation tool" of scientific data but also a key link between monitoring, analysis, decision-making, and public participation — it allows the invisible killers of air pollution to emerge. Let the "treatment process" of black foul waters be transparent, let the "health risks" of soil pollution be assessed, and let the "ecological costs" of climate change be felt. Through visual language to reduce cognitive thresholds, strengthen responsibility awareness, and guide the direction of action, environmental visualization is becoming a core supporting technology to promote "harmony between man and nature."

Chapter 10

Visualization Practices in Typical Interdisciplinary Areas of Science

10.1. Introduction: When Scientific Boundaries Melt Away in Visualization — The "Visual Connectivity" and "Catalyst for Innovation" of Interdisciplinary Exploration

At the forefront of contemporary science, the cross-cutting and fusion of disciplines has become the core driving force driving the expansion of knowledge boundaries — from the intersection of quantum physics and biology to quantum biology. It is difficult to solve the problem in science from the perspective of single subject. These interdisciplinary areas often involve multi-dimensional data (such as physical parameters, biological indicators, and socio-economic variables), heterogeneous data sources (such as laboratory measurements, simulation outputs, and field surveys), and complex system correlations (such as nonlinear coupling between variables and cross-scale effects). Their research process places a higher demand on cognitive tools: not only to accurately present data characteristics of a single discipline but also to reveal the mechanisms of interaction between multidisciplinary elements, helping researchers to break through the limitations of thinking in a single paradigm and discover new scientific patterns in cross-cutting areas (Zhang, *et al.*, 2023; Liu *et al.*, 2020).

Scientific visualization is just the "visual connector" and "innovative catalyst" in interdisciplinary studies, connecting different fields of knowledge and fusing multiple cognitive perspectives. It moves microscopic particles in physics through a unified visual language such as color coding, spatial mapping, and dynamic interaction. Heterogeneous data such as molecular structure dynamics in biology, group behavior patterns in sociology, and space-time distribution laws in economics are transformed into visual expressions that can be compared, linked, and explored, thus building a "cognitive bridge" between researchers in different disciplines. For example, in the cross-cutting research of brain science and artificial intelligence, visualization techniques can simultaneously present the electrical signal activity of neurons (bioelectric data) and the weight distribution of deep learning models (mathematical parameters), helping researchers understand the potential link between biological intelligence and artificial neural networks. In collaborative research between earth sciences and public health, visualization can overlay climate data (e.g., temperature and precipitation) with disease transmission data (e.g., case distribution and transmission rate) to reveal dynamic mechanisms of impact of environmental factors on public health (Kumar *et al.*, 2022; Hansen *et al.*, 2023).

Interdisciplinary visualization practices are unique in that they need to address both the challenges of "scientific complexity" and "cognitive diversity": On the one hand, data characteristics and research objectives vary significantly across disciplines (e.g., precision measurements in physics vs. social survey data). Microstructure in biology vs. macro-statistics in economics requires that visualization schemes must be flexible in adapting to the characteristics of multiple types of data (such as multi-scale spatial resolution, heterogeneous time–frequency, mixing of classification and quantitative variables). Interdisciplinary teams, on the other hand, have diverse backgrounds (e.g., scientists, engineers, and policy makers). Public participants also have different needs for visualization (e.g., researchers are concerned with the depth of regular findings, decision makers need to quickly capture key information, and the public expects intuitive presentation). Therefore, interdisciplinary visualization is not only the application of technical tools but also a set of tools. "Problem-oriented, With synergy at the core" Practical methodology — it requires starting from specific interdisciplinary research issues, integrating multidisciplinary knowledge frameworks, designing visualization strategies that balance scientific rigor and cognitive friendliness, and ultimately promoting the deep integration of knowledge and innovative breakthroughs in different fields (Wang *et al.*, 2022; Zhao *et al.*, 2023).

This chapter demonstrates how scientific visualization can function as a catalyst for multidisciplinary collaborative innovation through a series of practical examples of visualization across typical fields:

- The intersection of life sciences and physics (e.g., molecular dynamics simulations using cryo-electron microscopy, quantum mechanics-molecular mechanics combined calculations for protein folding processes), visualizing how the physical forces of microscopic particles (e.g., hydrogen bonds and van der Waals) and the structural evolution of biological macromolecules can be synchronized, helping to shed light on the molecular basis of life activities.
- Integration of earth and environmental sciences (e.g., climate change impacts on ecosystem service functioning, coupled analysis of geological hazards and socioeconomic risks) to visualize how climate data, ecological indicators, and human parameters can be integrated to support multidimensional decision-making for sustainable development.
- The synergy of computer science and biomedicine (e.g., intelligent diagnosis of medical images, virtual screening of drug molecules and target proteins), visualizing how high-dimensional feature data (such as pixel matrices of MRI images, energy scores of molecular docking) can be transformed into intuitive 3D models and interactive interfaces, helping doctors and researchers to quickly locate critical information.
- Extension of the social sciences and humanities (such as the analysis of spatial–temporal patterns of urban population movements and immersive displays of digital cultural heritage) visualizes how dynamic spatial charts and virtual reality technologies reveal the deep connection between human behavior and cultural values.

These cases not only demonstrate specific application scenarios for visualization techniques in interdisciplinary research (e.g., data integration, regular discovery, and knowledge dissemination), but they, moreover, reveal the core methodology of interdisciplinary visualization — problem-driven, based on multidisciplinary knowledge convergence, centered on user cognitive needs, breaking down disciplinary barriers, and inspiring innovative ideas through the uniformity of visual language and the flexibility of interaction design.

At a time when science is becoming increasingly integrated and complex, the practice of interdisciplinary visualization is becoming the key force driving knowledge innovation. It is not only a "visual assistant" for scientific discovery but also a "common language" for the dialogue between researchers in different fields, allowing the fragments of knowledge scattered across different disciplines to be tightly connected through visual bonds, ultimately putting together a more complete and profound scientific picture. The exploration of this chapter will open a window for readers to observe the frontier of interdisciplinary studies and provide practical enlightenment and innovative ideas for the application of scientific visualization in a broader field.

10.2. Physical Sciences and Engineering Visualization: Exploring the "Visual Decoder" of Matter and Energy

The core of research in the field of physical science and engineering is to reveal the basic structure of matter, the laws of interaction, and the mechanisms of energy conversion and transmission and to apply these scientific principles to the solution of practical engineering problems. Its research objectives span the micro (e.g., atoms and quarks), the meso (e.g., nanomaterials and nanodevices), and the macro (e.g., large mechanical equipment). Astro systems are multi-scale, encompassing both the exploration of fundamental physical laws (such as quantum mechanics, relativity, and electromagnetic theory) and engineering practices that translate scientific theories into practical technological applications (such as aerospace engineering, energy engineering, and electronic information engineering). In doing so, researchers and engineers face a number of challenges: Physical phenomena in the micro-world are often difficult to observe directly (such as the trajectories of electrons, the superposition and entanglement of quantum states), and macroscopic engineering systems have a high degree of complexity and dynamicity (e.g., multi-component aero engines working together, large-scale energy transmission in power networks, and the engineering design and optimization process requires in-depth analysis and intuitive presentation of large amounts of data (such as structural stress, fluid flow, and electromagnetic field distribution). Scientific visualization technology provides a key support to solve these challenges. It can transform abstract data in physical science and engineering (such as physical field distribution, structural mechanical parameters, and energy conversion efficiency) into intuitive visual representations (such as graphics, images, dynamic simulations, or interactive interfaces) that assist researchers in theoretical verification and mechanism discovery through multi-scale presentations, dynamic process presentations, and interactive exploration and support engineers in engineering design, performance evaluation, and optimization decisions (Liu *et al.*, 2020; NASA, 2022).

The essence of physical science and engineering visualization is to take the "intangible" properties of matter and energy (such as the spatial distribution of electromagnetic fields, the probability distribution of quantum states) and the "dynamic" behavior of complex systems (such as mechanical components' motion processes). The temporal variation of fluid flow is transformed into "tangible" visual information through visual metaphors (e.g., color mapping of electric field intensity). Multi-scale integration (from micro-particles to macro-equipment system-level correlation) and dynamic presentation (such as the evolution of physical processes and engineering system state in service) help users understand physical laws, optimize engineering design, and promote technological innovation. Its application scenarios span basic physics research (such as particle physics experimental data analysis), engineering design and manufacturing (such as the structural optimization of machinery and equipment), energy development and utilization (such as wind power generation aerodynamic performance analysis, nuclear reactor thermal hydraulic simulation), and aerospace (such as aircraft aerodynamic shape design and flight path simulation) (Zhao *et al.*, 2023).

This section focuses on the application scenarios of physical science and engineering visualization. Three core content areas are supported by digital media technology and typical practice cases, and the system analyzes how it can become a "visual decoder" for exploring matter and energy through multi-scale visualization, dynamic process simulation, and interactive exploration.

10.2.1. *Application Scenarios for Visualization in Physical Science and Engineering: Multi-scale Requirements from Micro-Systems to Macro-Engineering*

Physical science and engineering research covers multiple scales and processes, and the requirements for visualization in different scenarios have commonalities (such as accuracy and intuitiveness), but they also have different focuses because of the different subjects and objectives. The following are typical application scenarios and their core requirements:

1. Microscopic Physics Visualization: Uncovering the mechanisms underlying the microscopic world
Situation: To study the structure of atoms, molecules, electrons, and other microscopic particles, interactions, and quantum properties (such as distribution of atomic orbits, shape of electron clouds, and quantum tunneling effects), as well as physical properties of microscopic materials (such as nanomaterials and semiconductor materials) (such as conductivity, magnetism, and optical properties).

Core requirements: These demonstrate the spatial distribution and motions of microscopic particles (such as the probability distribution of electrons outside the nucleus and the vibration and rotation of atoms in molecules), the mechanisms of microscopic interactions (e.g., chemical bond formation and breakdown, quantum state superposition and entanglement), and the relationships between the physical properties of microscopic materials and their microstructure (e.g., nanomaterial size effects and surface effects).

2. Macrophysics and Engineering Visualization: Supporting Engineering Design and Performance Optimization
Scenarios involving mechanical engineering (such as structural design and performance optimization of automotive engines and aero engines), civil engineering (such as bridges and high-rise buildings), electronic information engineering (such as electronic chip thermal management and circuit board EMC design), energy engineering (such as wind power generation aerodynamic performance and nuclear reactor thermal hydraulic properties), and many other engineering fields.

Core requirements: These present the mechanical properties of an engineering structure (e.g., stress, strain, and displacement distribution), fluid flow characteristics (e.g., speed), electric field intensity, magnetic field strength, electromagnetic wave propagation, power conversion (such as heat to mechanical energy and electric power to light energy), and project state in service (such as engine speed, power output, and fuel consumption).

3. Physical Process Visualization: Dynamically Presenting the Evolutionary Process of Physical Phenomena

Study the dynamics of physical phenomena, such as mechanical vibration, fluid flow, electromagnetic induction, and thermal conduction, as well as the evolution of these phenomena under different conditions (such as the changes in fluid flow patterns under varying Reynolds numbers or differences in heat conduction rates under different temperature gradients).

Core requirements: These demonstrate the temporal evolution of physical processes (e.g., the velocity field of fluid flow over time and the dynamic distribution of electromagnetic field strength), the interaction of process parameters (e.g., the coupling relationship between fluid velocity and pressure, and the correlation between temperature and thermal flow density), and the spatial distribution characteristics of key physical quantities (e.g., the vortex structure in fluids and the equation-face distribution in electromagnetic fields).

4. Experimental and analog data visualization: Bridging the gap between theory and practice

Scenarios: Physical science and engineering research relies on large amounts of experimental data (such as detector records in particle physics experiments and mechanical and electrical measurements in the measurement of material properties) and numerical analog data (such as structural stress in finite element analysis and fluid flow in computational fluid dynamics).

Key requirements: These convert experimental data (such as discrete data points collected by sensors) and analog data (such as data generated by numerical calculations) into intuitive and visual representations. Through data mining and analysis, new physics rules and project optimizing directions are found.

10.2.2. *The Core Support of Digital Media Technologies: From Data Acquisition to Interactive Exploration*

The efficient realization of physical science and engineering visualizations relies on several key capabilities of digital media technologies that underpin multi-scale data integration, dynamic process simulation, and interactive analysis:

1. Multi-source data acquisition and processing technique

- **Experimental data acquisition:** Various sensors (e.g., mechanical sensors to measure structural stress, temperature sensors to measure temperature distribution, pressure sensors to measure fluid pressure) and detectors (e.g., particle detectors in particle physics experiments and telescopes in astronomical observations) obtain a variety of parameter data from physical processes. These data usually have high accuracy and high temporal and spatial resolution but can have noise and data missing issues, requiring data cleaning, calibration, and inserted value processing.
- **Analog data generation:** This is done using finite element analysis (FEA), CFD, and electromagnetic modeling (e.g., node and unit data in structural mechanics analysis, and velocity and pressure field data in fluid simulation). These data need format conversion, dimension reduction, and data compression to improve the efficiency of visualization.

2. Multi-scale visualization techniques

- **Micro-scale visualization:** For microscopic particles and materials, using atomic-scale modeling (such as quantum mechanical calculations to obtain the position of atoms and the distribution of electron clouds) and molecular dynamics simulations (showing the movement and interaction of molecules) also presents micro-structure and interaction mechanisms through 3D visualization techniques (such as ball stick models and electron cloud density maps) to help researchers understand the physical laws of the micro-world.
- **Macro-scale visualization:** For engineering structures and large physical systems, use three-dimensional geometric modeling (such as creating a three-dimensional model of mechanical equipment based on CAD design drawings, and constructing terrain models of civil engineering structures based on terrain data) and finite element mesh partitioning (dividing the structure or fluid region into multiple small units for use in mechanical analysis and fluid simulation). Employ surface rendering (to display the appearance and shape of the structure) and volume rendering (to present the distribution of internal physical fields, such as stress distributions and electromagnetic field distributions) techniques to provide an intuitive representation of key information related to macro-scale physical and engineering problems.

- **Cross-scale correlation visualization:** This include linking and integrating information at the micro and macro scales. For example, when studying the macroscopic properties of nanomaterials, their microstructure (such as atomic arrangement and grain size) and macroscopic mechanical properties (such as strength and toughness) can be correlated and visualized, revealing the mechanism of influence of microstructure on macroscopic properties and providing guidance for materials design and engineering applications.

3. Dynamic visualization and simulation techniques

- **Dynamic simulation of physical processes:** This is the simulation of the dynamic changes of physical processes based on numerical computational methods, such as the analysis of time sequence of mechanical vibration. The dynamic process of electromagnetic induction (showing the dynamic distribution and variation of electric and magnetic field strength), etc., through dynamic visualization technology (such as animation and video), intuitive presentation of the evolution of physical phenomena, to help users understand the physical laws and process mechanism.
- **Real-time data visualization:** This is for real-time data from experimental processes and engineering operations (such as structural stress data collected by sensors in real time and operational parameters monitored by engines in real-time). Through the real-time visualization technology (such as dynamic chart and real-time monitor interface), the data can be transformed into intuitive visual information, which can help researchers and engineers to grasp the physical process and the running state in service, discover the abnormal situation, and make decision in time.

4. Interaction and Immersive Technologies

- **Interactive exploration:** Users can manipulate visual objects, such as rotating, using interactive devices such as a mouse or keyboard, scaling and flattening 3D models, selecting different physical parameters for visualization, viewing details of specific areas or objects (such as the stress value of a certain point in a structure and the flow rate of a fluid at a given point), and understanding physical problems and the properties of an engineering system through interactive analysis.
- **Virtual reality (VR)/augmented reality (AR) technology:** This includes engineering design and training scenarios. Through VR

devices, users can immerse themselves in a virtual engineering environment (such as the internal structure of an aero engine and the construction site of a large building), visually observe the details of the engineering structure and the physical processes occurring. Through AR technology, virtual physical information (such as electromagnetic field distribution and stress cloud maps) is superimposed on real engineering objects (such as machinery and buildings), assisting engineers in field inspection, maintenance, and design optimization, improving work efficiency and decision accuracy.

10.2.3. *Typical Practice Cases: Visualization Technologies Enable Innovation in Physical Science and Engineering*

Case 1: Air engine aerodynamics visualization analysis — a key tool for optimizing design

The scenario: The aircraft engine is the "heart" of the aircraft, and its aerodynamic performance (e.g., inlet efficiency, compression efficiency, and combustion efficiency) directly affects the performance and fuel economy of the jet. In the design and development of aero engines, in-depth analysis and optimization of aerodynamic form (such as blade shape, inlet structure) and airflow characteristics (such as speed distribution, pressure distribution, turbulence structure) are required to improve the efficiency and performance of the engine.

The visualization scheme:

- **3D aerodynamic shape modeling and visualization:** Based on CAD design drawings, construct 3D geometric models of various components of aero engines (such as compressor blades, turbine blades, and inlet channels), visually display the engine's shape structure through surface rendering technology, and assist engineers in shape design and optimization.
- **Dynamic simulation and visualization of airflow:** Using computational fluid mechanics (CFD) methods, the numerical simulation of airflow inside the air launch motor, the grid data such as speed, pressure, and intensity of turbulence are generated. Visual techniques such as pressure cloud maps (representing the spatial distribution of pressure) and turbulence visualizations (such as dynamic displays of vortex structures) visualize the flow process and characteristics of air flow inside the engine, helping engineers analyze loss of air flow and optimize direction.

- **Key parameter correlation analysis and visualization:** Connecting airflow parameters (such as velocity and pressure) to engine performance parameters (e.g., efficiency and thrust) performs correlation analysis, demonstrates the patterns of changes in engine performance under different aerodynamic shapes and flow conditions through multivariable visualization charts (such as scatter plots and heat plots), and assists engineers in determining the optimal combination of design parameters to improve the overall performance of the engine.
- **Interactive exploration and optimization of design:** Engineers can use interactive interfaces. It can shorten the research and development cycle of the engine and reduce the research and development costs.

Application of Digital Media Technology: This case integrates CAD modeling technology, CFD numerical simulation technology, and 3D visualization technology. The rapid processing and visual rendering of large-scale grid data through a high-performance computing platform, utilizing an interactive interface and dynamic visualization technology, provides engineers with an intuitive and efficient design and optimization platform, significantly improving the design quality and performance of aircraft engines (NASA, 2023; Kumar *et al.*, 2022).

Case 2: Experimental Data Visualization of Quantum Physics — Exploring the Mysteries of the Microworld

Situation: Quantum physics research involves the quantum properties of microscopic particles (such as superposition of quantum states, quantum entanglement, and quantum tunneling). These phenomena are difficult to fully explain by classical physics theory and need to be observed and studied through experiments. In quantum physics experiments, detectors record large amounts of microscopic particle data (such as the location of the particles, energy, and spin states) — these data contain a wealth of information about the quantum world but are often characterized by high dimensions, complexity, and uncertainty, requiring visual analysis to reveal the physical laws within.

The visualization scheme:

- **Probability distribution visualization of quantum states:** For the quantum positions of quantum systems (such as the electron orbit of atoms and the state of qubits), the probability distribution of the quantum state (such as the probability density distribution of electrons in space outside the nucleus) is computed by quantum mechanics. The probability of quantum bits in different states, using three-dimensional visualization techniques (such as electron cloud density maps and probability spheres) to display the spatial distribution characteristics of quantum states, helps researchers understand the basic properties of quantum systems.

- **Multi-dimensional presentation of experimental data:** Visualize a large amount of data recorded by detectors in quantum physics experiments (such as the location, energy, and time information of particles) in multiple dimensions, such as showing the location and energy distribution of particles through a three-dimensional scattering diagram). Time series graphs show the changes of particle states over time, heat graphs show probabilities of quantum phenomena under different experimental conditions, and multivariate association analysis reveals the relationship between quantum phenomena and experimental parameters.
- **Quantum entanglement and quantum tunneling visualization:** For quantum entanglement, demonstrate the nonlocality and correlation of quantum entanglement by visualizing the state correlation of two or more entangled particles (such as the evolution of the entangled state of two qubits over time). For quantum tunneling phenomena, by simulating the process of particle movement near the potential barrier (such as the distribution and change of the particle's wave function inside and outside the potential shield), the probability and mechanism of particle passage through the potential defense can be visualized, helping researchers to understand the strange phenomena of the quantum world in depth.
- **Interactive data analysis and exploration:** Researchers can select different subsets of experimental data (such as data under different experimental conditions) through an interactive interface. Data mining and pattern recognition are carried out to discover the physical rules and new quantum phenomena hidden in the data.

Digital media technology applications: This case utilizes quantum mechanics computing software, data processing and analysis tools, and 3D visualization technologies, through interactive interfaces and multi-dimensional visualizations, which provides researchers with a powerful tool to explore the mysteries of the quantum world, promoting research and development in the field of quantum physics (Zhang *et al.*, 2021; Eyring *et al.*, 2021).

Case 3: Visual monitoring and optimization of battery thermal management system for new energy resources vehicles — Key to ensure safety and performance

Scenario: One of the core components of new energy resources vehicles is lithium-ion battery pack, whose performance and safety are closely related to battery temperature. The battery will generate a lot of heat during charging and discharging, and if the heat cannot be dissipated in a timely and effective manner, it will cause the battery temperature to be too high, which not only affects the battery's service life and performance but may also trigger safety problems such as heat loss. Therefore, real-time monitoring and optimization of the battery thermal management system (including coolant flow, temperature distribution, and heat exchange processes) is essential.

The visualization scheme:

- **Battery temperature field visualization:** By placing multiple temperature sensors inside the battery pack, the temperature data of each battery monomer is collected in real time, and three-dimensional visualization technology displays the temperature data in the form of color gradients (e.g., blue indicates low temperature, and red indicates high temperature) on the three-dimensional model of the battery pack, visualizing the temperature distribution within the battery pack. For example, when a vehicle is charged rapidly, the visualization interface can clearly show areas where certain battery monomers experience local temperature increases due to high charging current, helping engineers quickly locate potential high-temperature risk points.
- **Visualization of coolant flow and heat exchange:** Combining coolant flow sensor and pressure sensor data and employing computational fluid dynamics (CFD) methods simulate the flow of coolant in the battery pack cooling pipe, including information such as flow rate, pressure, and direction, and visualize this information superimposed with temperature field data. The flow trajectory of the coolant is depicted in flow lines, the temperature of the coolant is depicted in different colors, and the efficiency of the heat exchange is visually demonstrated by how the coolant takes away the heat generated by the battery.

For example, when the coolant flow rate is low or the pipeline is blocked, the visual interface can show that the cooling effect is poor, and the temperature increases in the corresponding area.

- **Dynamic simulation and optimization of heat management strategies:** Based on different heat management strategies (such as changing coolant flow and adjusting the rotation speed of the cooling fan), optimize the cooling pipe layout and demonstrate the impact of these strategies on the battery temperature distribution and heat exchange efficiency in real time through numerical simulation and visualization techniques. Engineers can adjust policy parameters (such as increasing coolant flow or increasing the speed of the cooling fan) through an interactive interface. Observe trends in battery temperature to choose the optimal thermal management strategy to ensure the battery operates within the safe temperature range while improving the overall performance and service life of the battery.
- **Fault warning and diagnostic visualization:** Establish a fault warning model through real-time monitoring and analysis of battery temperature data. When the temperature of a battery monomer increases abnormally or the temperature distribution becomes uneven, the visual interface will issue an early warning signal in time. By highlighting the abnormal area and the relevant sensor data, the assistant engineer can quickly diagnose the cause of the fault (such as coolant leakage and battery internal short circuit) and take appropriate measures to repair, to ensure the safe operation of new energy resources vehicles.

Digital media technology application: The case integrates sensor network technology (collecting temperature, flow, and data in real time), CFD numerical simulation technology (simulation of coolant flow and heat exchange process), and 3D visualization technology (showing temperature field and flow process). Through real-time data updating and interactive operation, it provides a comprehensive and intuitive solution for the monitoring, optimization, and fault diagnosis of the new energy resources vehicle battery thermal management system, which helps improve the safety and reliability of new energy vehicles (Gupta *et al.*, 2021; Wang *et al.*, 2022).

Case 4: Visual Study of Superconducting Materials Microstructure and Superconducting Mechanisms — The Key to Unraveling Superconductivity

Scenarios: Superconducting materials have zero resistance under specific low-temperature conditions and have tremendous potential for applications such as efficient power transmission and strong magnetic field generation. However, the superconductivity mechanism has not been fully defined, and the relationship between its microstructure (such as crystal structure and electron state distribution) and superconductivity performance is the focus and difficulty of research. Through visualization technology, the microscopic structural characteristics of superconducting materials can be studied in depth, exploring superconducting mechanisms and providing theoretical support for the development of new superconducting materials.

The visualization scheme:

- **Modeling and visualization of the microstructure of superconducting materials:** These include the use of X-ray diffraction, electron microscope, and other techniques to obtain crystal structure data of superconducting materials (e.g., atomic positions, the imperfection of crystal lattice, and the arrangement of atoms are directly displayed). For example, for copper-based superconducting materials, the atomic arrangement and interlayer structure of the copper–oxygen surface layer can be clearly shown, helping researchers understand its effects on superconducting properties.
- **Visualization of electronic state distributions and superconducting energy gaps:** The density of states and the superconducting energy gap in superconducting materials are calculated by using the density functional theory (DFT). The density of states and the range of superconducting energy gaps are represented by color gradients and equal-value surfaces, respectively. For example, the size and distribution of superconducting energy gaps are closely related to superconductivity critical temperatures, and the differences in superconducting power gaps between different superconducting materials can be observed through visualization, providing clues to the search for high critical temperature superconducting material.
- **Dynamic simulation and visualization of superconductivity mechanism:** By simulating microscopic processes such as electron–acoustic interaction (a key process in superconductivity), the formation and motion of Cooper pairs, animation and dynamic visualization techniques are used to demonstrate the microscopic nature of superconducting phenomena. For example, show how electrons interact with

wafer vibrations (phonons) to form Cooper pairs, and how Cooper pairs move unimpeded through the wafer to achieve zero-resistance superconductivity. This dynamic visualization helps researchers to gain a deeper understanding of superconductivity mechanisms and provides intuitive support for theoretical research.

- **Material performance correlation analysis and visualization:** Microstructure characteristics of superconducting materials (such as crystal defects and electron state distribution), superconductivity mechanism (such as the formation of Cooper pairs), and superconductivity performance (such as critical temperature and critical current density) are analyzed in relation to each other, and the effects of different microscopic structural parameters on superconductivity properties are shown through multivariable visualization graphs (such as scattering graphs and heat graphs). Scientists can adjust microstructure parameters (such as changing lattice parameters and introducing specific crystal defects) through an interactive interface to observe changes in superconducting performance, thereby providing guidance for optimizing the design of superconducting materials.

Digital media technology applications: The case utilizes advanced materials characterization techniques (obtaining microstructure data), quantum mechanical computing software (simulation of electron states and superconductivity mechanisms), and 3D visualization techniques (showcasing microstructure and superconducting processes). Through multi-dimensional information integration and intuitive visualization, the research of superconducting materials is provided with a powerful tool that helps advance the development and application of superconducting science (Zhao *et al.*, 2023).

The Common Value and Future Outlook of Visual Cases in Physical Sciences and Engineering

The above four cases (thermal management of new energy resources automobile battery and research on microstructure of superconducting material) and the original case (aeroengine aerodynamic performance analysis and quantum physics experiment data visualization) together reflect the core value of visualization technology in the field of physical science and engineering:

- **Multi-scale correlation:** It can analyze not only the micro-level (such as quantum state and electronic state distribution of superconducting materials) but also the macro-level (such as aircraft engines and new energy resources vehicle batteries).

- **Dynamic process presentation:** Through dynamic simulation and real-time monitoring (such as air flow, battery temperature changes, and microprocesses of superconductivity mechanism), the time-series evolution of physical phenomena is revealed, and researchers are assisted in understanding the nature of dynamic processes.
- **Interactive exploration and optimization:** Interactive exploration and optimization enable researchers and engineers to adjust parameters (such as engine design parameters, thermal management strategies, and microstructure parameters) in real time through an interactive interface, observe changes in results, quickly find optimal solutions, and drive engineering optimization and innovation.
- **Interdisciplinary integration:** The knowledge and technologies of physics, engineering, computer science, and other disciplines are combined to transform complex physical information into intuitive visual expression through digital media technology, promoting communication and cooperation between different disciplines.

In the future, with artificial intelligence (such as automatically identifying key features in physical phenomena and predicting material properties), virtual reality (such as immersive exploration of the microscopic physical world), and digital twins (such as building virtual models of physical systems), with the further development of technologies such as real-time synchronization and prediction, physical science and engineering visualization will evolve toward a more intelligent, immersive, and precise direction, providing stronger support for exploring the mysteries of matter and energy and driving scientific and technological innovation.

Core values of visualization in physical science and engineering

The core value of physical science and engineering visualization is its ability to translate the complex properties of matter and energy and the dynamic changes of physical processes into intuitive visual information. Through multi-scale demonstrations, dynamic process simulations, and interactive exploration, researchers and engineers are helped to better understand physical laws, optimize engineering design, and drive technological innovation. It is a bridge between physical theory and engineering practice and an important tool for exploring the microscopic world and solving macroscopic engineering problems. With the continuous development of digital media technologies (such as AI-assisted data analysis, widespread use of virtual reality, and augmented reality), physical science and engineering visualization will play an even more important role in the future, providing stronger support for humans to understand nature and transform the world. In the following chapters, we further explore the cross-disciplinary appeal of scientific visualization in other typical scientific fields, such as materials science and energy science.

10.3. Cross-Visualization of Life Sciences and Physics: Exploring the Physical Nature of Life Phenomena

Cross-cutting fields of life sciences and physics (e.g., biophysics, biomechanics, and quantum biology) work to uncover "basic physical laws governing life activities" — from the dynamics of protein folding to the electrical properties of ion channels in cell membranes. From the magnetic induction mechanism of bird migration to the transmission of electrical signals in cranial nerves, the answers to these questions are often hidden in the deep integration of physical principles and life processes. However, the objects of such research (e.g., molecular conformation changes and micro-energy transfer) have the characteristics of extremely small scales (nano-microscale), extremely fast processes (picosecond-millisecond), and extremely complex mechanisms (multi-physics field coupling), which are difficult to visualize in traditional text or single-dimensional graphs. To solve the problem in science, it is necessary to study the relationship between physics and biology.

10.3.1. *Core Visualization Needs and Challenges in Cross-Cutting Areas*

Cross-cutting research in life sciences and physics often focuses on "physical processes in living systems" or "the effects of physical fields on living activities," and its visualization needs to meet the following requirements:

- **Multi-scale co-ordination:** These are both to demonstrate macroscopic life phenomena (such as the mechanical field distribution when the heart pumps blood) and to characterize microscopic physical mechanisms (such as opening and closing dynamics of calcium channels within heart muscle cells).

- **Physical–biological parameter association:** This is the dynamic association of physical quantities (such as force, electric field intensity, and temperature) with biological indicators (such as cell activity, protein function, organ function) (such as "myocardial contractility and calcium concentration correlation").
- **Dynamic restoration:** This is to capture the cause-and-effect relationship of rapidly changing physical-life events, such as millisecond conduction of electrical pulses of neuron and step motion of molecular motor.

Challenges include heterogeneity of data sources (e.g., incompatibility of biological and physical analog data formats), microscopic mechanistic abstractions (such as the role of quantum effects in photosynthesis), and interdisciplinary terminological barriers (such as differences in the description of "energy transfer" between biologists and physicists).

10.3.2. *Typical Visualization Scenarios and Technical Strategies*

Scenario 1: Physical-life coupling at the molecular level — using protein folding and enzyme catalysis as an example

Background of the study: The function of proteins depends on their three-dimensional structure, and the folding process is driven by thermodynamics (the minimum principle of free energy) and dynamics (interatomic interactions). Enzyme-catalyzed reaction can accelerate metabolism by reducing reaction activation energy (physicochemical principle).

Objective of visualization: To dynamically demonstrate the folding path of a protein from a linear amino acid chain to a functional conformation, or the process of conformation change and energy change when an enzyme binds to substrate.

Technology implementation:

- **Data sources:** These include Molecular Dynamics (MD) simulation data (computer simulation of forces and motions between atoms to generate snapshots of molecular composition per skin second) and

X-ray crystallography/cryo-electron analysis of experimental structures (static high-precision 3D models).

- **Visual design:**

Folded Path Animation: Mark amino acid residues with different colors (e.g., hydrophobic cores in blue, hydrophilic surfaces in red) to dynamically illustrate how chain-like molecules gradually fold into specific conformations through hydrogen bonds and van der Waals forces (e.g., the assembly of β-sheet layers with α-helices), and label key intermediate states (e.g., the hydrophobic aggregation in "glassy state").

Mapping free energy (longitudinal axis) to conformational characteristics (transverse axis, such as critical dihedral angle) into a three-dimensional surface, using a "ball" to represent the state of a molecule at various moments, animated the process of rolling down the lowest energy path to a stable cone (intuitively illustrating the nature of "thermodynamic driving").

Catalytic details: These are obtained by comparing conformational changes (e.g., shifts of amino acid residues at the active site) to "before/after" substrate binding, in conjunction with the reaction axis (the horizontal axis is the reaction process, and the vertical axis is the change of free energy), the value of the reduction of the activation energy (e.g., "from 80 kJ/mol to 40 kJ/mol").

Harvard University's "Molecular Movies" library contains hundreds of visual animations of protein folding and enzyme catalysis. Users can interactively adjust the view angle (e.g., focusing on tiny displacements of catalyst residues) and view corresponding physical parameters (e.g., binding energy, reaction rate constant).

Scenario 2: Physical field regulation at the cellular level — using cell mechanics and electrophysiology as examples

Context of research: Cell proliferation, migration, differentiation, and other behaviors are significantly affected by physical fields (such as mechanical stress, electric field gradient, and fluid shear). The synchronized contraction of cardiac myocytes depends on the transmission of electrical signals (Action potential), and the information processing of neuronal networks is based on the opening and closing of ion channels (Na/K electrochemical gradients).

Objective of visualization: To reveal how physical fields regulate cellular behavior or how cells respond to and transform physical signals into biological functions (e.g., "How mechanical forces activate the YAP/TAZ signaling pathway within cells").

Technology implementation:

- **Data sources:** These include cellular mechanical properties (e.g., stiffness, patch clamp technique, action potential time history, and ionic current peak).
- **Visual design:**

Mechanical field map: The distribution of the elastic modulus of the extra cellular matrix (ECM) is shown with a pseudo-color overlay (e.g., the hard areas in the tumor microenvironment are red, and soft areas are blue), superimpose cell morphology (label cell skeleton by fluorescence staining), and animation demonstrates how cells migrate to high hardness areas ("hardening" behavior).

Action potential (AP) recordings from a population of myocardial cells are translated into "electrodynamic" animation (highlighted in red to represent depolarized regions and blue for the repolarization area) and mark the conduction velocity (such as "normal ventricular conduction velocity is 0.5–1 m/s, the fine area decreased to 0.1 m/s leads to cardiac rhythm").

Ion channel mechanism: The "open-inactivated-closed" state cycle of the sodium ion channel (the trajectory of the internal flow of the ion when combined with electric field forces) is demonstrated in a 3D model, with membrane potential changes (e.g., "channel opens when depolarizing from −70 mV to + 30 mV").

Typical example: Stanford University's Cell Physics Lab visualization platform shows morphological changes in endothelial cells under the force of blood flow pruning (0.1–10 Pa) (from polygonal spread to slender alignment). The quantitative variation of nitric oxide (NO, vasodilation factor) was correlated (the higher the shear force, the more NO released → the more significant vasodilation).

Scenario 3: Biophysical phenomena at the macro level — animal motion and sensory physics as examples

Background: Bird migration depends on the perception of the geomagnetic field (quantum effects of magnetic receptor proteins), fish cluster movement follows the laws of fluid mechanics (reduced drag and energy consumption), and gait optimization in human running involves the balance of mechanics and energy metabolism.

Objective of visualization: To analyze the physical drivers of macroscopic biological behavior or how the physical environment shapes biological adaptive evolution (e.g., "How magnetite particles in the retinas of migratory birds sense the direction of the geomagnetic field").

Technology implementation:

- **Data sources:** These include field tracking data (temporal and spatial correlation of GPS bird migration routes with geomagnetic field strength/direction), animal motion video captured by high-speed cameras (analysis of gait cycles, group collaboration), and numerical simulation (e.g., computational fluid mechanics CFD simulation of eddy current distribution of fish swimming).
- **Visual design:**
  Magnetic induction simulation: This converts the quantum spin state of a magnetic receptor protein in the bird's retina (affected by the geomagnetic field) into a "virtual compass" animation (red arrows represent the perceived geomagnetic field direction), mapping the range of distribution of key proteins such as CRY4.
  Group motion flow fields: Particle systems simulate the tail flow (whirlpool structure) when a fish swims and indicate the interaction between individuals through a velocity vector arrow (e.g., "When the leader fish slows down, the rear fish automatically adjust their spacing through a pressure gradient").
  Human gait mechanics: A pressure sensor array collects the pressure distribution of various areas of the soles of the feet during running (e.g., peak pressures on the forefoot and heel), combined with kinetic energy consumption data (calories consumed per kilometer), and labeled optimal gait patterns (e.g., "the grounding of the front paw reduces the braking effect and reduces energy consumption by 15%").
  Typical example: This includes the visualization project of the physics of animal movement at the Max Planck Institute in Germany shows albatrosses using "dynamic aerial" techniques (using the rising and falling zones of sea air flow) to fly long distances (almost without flapping their wings). The mystery of "physical environment enabling biological

motion" is presented intuitively through the layering of the altitude field of the air flow (warm = rising air flow, cold = falling air flow) and the superposition of the albatross's flight trajectory.

10.3.3. *Technical Tools and Design Principles for Cross-Visualization*

Commonly used technical tools:
Molecular/cellular level: PyMOL/VMD (visualization of molecular structure), ImageJ/Fiji (microscopic image processing), and COMSOL Multiphysics (coupled physics).

Macro-level: ArcGIS (geospatial and physical superposition), MATLAB/Simulink (dynamic modeling and simulation), and Blender (three-dimensional animation rendering).

Core design principles:

- **Physical accuracy:** This ensures that the parameters in the visualization (e.g., force/electric field units, reaction rate constants) are consistent with experimental or analog data and avoid distorting the nature of science for visual effects.
- **Bio-correlation:** This directly links physical processes to life functions (e.g., "calcium ion concentration change → muscle contractile strength") and avoids isolating physical phenomena.
- **Interdisciplinary readability:** This uses a unified system of terms (e.g., "free energy" rather than "energy") and provides biologists with different perspectives (e.g., "functional outcomes" for biologists and "institutional details").

Visualization makes the "life-physics" dialogue more intuitive. Cross-visualization of life science and physics is essentially a bridge between the laws of microphysics and macroscopic life phenomena through visual language — it makes the "energy valley" of protein folding visible. A "synchronous dance" of electrical signals in heart muscle cells can be sensed, and a "quantum compass" that allows birds to sense magnetic fields can be tracked. Such visualizations not only advance basic research in cross-cutting fields (such as uncovering the physical constraints of life's origin and designing bionic materials) but also enable the public to understand the profound proposition that life is not only chemical but also physical. In the future, with the development of technologies such as cryo-electroscopes and quantum sensing, cross-visualization of life and physics will further expand the boundaries of human understanding of the nature of life.

10.4. Integration Visualization of Earth and Environmental Science: Analytical of the Dynamic Coupling of Human-Terrestrial Systems

Geosciences focus on "the structure of systems" (such as geological structure, general atmospheric circulation, and hydrological cycle), emphasizing the long time scale (millions to thousands of years), the objective law. Environmental science focuses more on "interactions between human activities and the natural environment" (e.g., pollution emissions, ecological destruction, and resource depletion) and focuses on dynamic responses at short time scales (tens of years). However, as the study of global change deepens, the two are moving from parallel to integrated — How, e.g., does geological activity, such as volcanic eruptions, affect regional climate and environmental quality? How can human activities, such as large-scale mining, affect geological stability by altering the circulation of material on the surface? Such problems need to integrate both the natural processes of the Earth system and the social drivers of the environmental system, and visualization techniques are a key tool for cracking this "natural-human coupling complexity."

10.4.1. *Core Visualization Needs and Challenges in the Area of Integration*

Cross-cutting research in Earth and environmental sciences requires simultaneous mapping of the objective laws of natural processes and the dynamic interference of human activities, and its visualization faces a threefold challenge:

- **Multi-scale nesting:** This is not only to present the global scale of plate movement (million years scale) or general atmospheric

circulation model (decade scale) but also to focus on the regional scale of mining (annual scale) or urban heat island effect (daily scale).
- **Process-coupling:** The flow of material through the circulation of material (e.g., nitrogen and phosphorus (N and P)) is related to energy flow (such as solar energy input), and the hidden feedback mechanism (such as "vegetation destruction → water loss and soil erosion → increased river sediment → reservoir siltation → increased flood risk") should be revealed.

Heterogeneous data: Most of the data come from remote sensing satellites (low spatial resolution but wide coverage). For geological drilling (high precision but point-wise distribution), environmental science data consist of ground monitoring (e.g., air quality stations and water quality sensors) and social statistics (e.g., population density and industry emissions inventories), with significant differences in format and time resolution.

Its core objective is to create a visual framework for "natural-human synergy." Integrating geological structures, climate patterns, human activities (e.g., urbanization and industrial emissions), and ecological responses (e.g., changes in biodiversity and environmental pollution) into dynamically linked wholes helps researchers identify key coupling nodes and predict system evolution trends.

10.4.2. *Typical Visualization Scenarios and Technical Strategies*

Scenario 1: Linkages between geological activity and regional environmental quality — using volcanic eruption as an example
Background: Volcanic eruption is a typical internal process of the Earth (magmatic activity and crustal movement), but it will directly release a large amount of sulfide (such as SO_2). Volcanic ash (particulate matter PM10/PM2.5) and greenhouse gases (e.g., CO_2) contribute to regional air quality deterioration and short-term global cooling ("volcanic winter") while affecting soil fertility and water resources (e.g., acid rain).

Visualization objective: To dynamically demonstrate the natural processes of volcanic eruptions (rise of magma → eruption → evacuation) and their multidimensional effects on environmental systems (atmospheric composition, climate, and ecosystems).

Technology implementation:

- **Data sources:** These include geological monitoring data (seismic waves monitoring for precursors of volcanic activity and GPS measurement of crustal deformation), satellite remote sensing (such as MODIS monitoring the spatial distribution and height of volcanic ash clouds), meteorological model (WRF simulation of ejecta dispersion path), and environmental monitor station (SO, PM2.5 concentration real-time data).
- **Visual design:**
  - **Three-dimensional simulation of the eruption process:** This visualizes the path of magma from the mantle to the surface through fluid dynamics models (such as MAGMA-CFD) (red high-temperature magma flow) and the stress distribution extended by crustal cracks (pseudo-color gradients show stress concentrated areas).
  - **Eruption spread and environmental response:** Particle systems simulate the spread of ash (gray particles) and SO_2 (yellow gas) with a combination of wind direction and altitude layer data, dynamically labeling affected areas (e.g., "ash cloud cover radius 500 km, air no-flight area") and simultaneous display of changes in environmental indicators (such as "3 days after the eruption, the concentration of PM2.5 in surrounding cities soared from 30 $\mu g/m^3$ to 800 $\mu g/m^3$, and the concentration of SO_2 exceeded the national standard by 10 times").
  - **Long-term climate effects:** These are through comparative simulations (with/without eruptions), showing a global temperature change curve (e.g., "Average Northern Hemisphere temperature decreases by 0.5°C within 1 year of an eruption") and correlating ecosystem responses (e.g., "Plant growth season delayed by 2 weeks at high latitudes").

In the United States Geological Survey (USGS) "Volcanic Hazard Visualization System" for Kilauea Volcano in Hawaii, the eruption site location, lava flow velocity (meters to hundreds of meters per day), and ash spread forecasts are updated in real time, and the public can use an interactive map to see if their location is in a hazardous area, which environmental authorities can use to initiate air quality emergency responses.

Scenario 2: Changes in surface processes driven by human activity — mine mining as an example

Background: Open mining is the direct transformation of the Earth's surface by humans (removal of rock). However, it will lead to a series of environmental issues: land subsidence (geological disasters), waste accumulation (soil pollution and water loss and soil erosion), heavy metal leaching (groundwater pollution), and vegetation destruction (biodiversity loss).

Objective of visualization: To reveal the temporal and spatial correlations between the natural processes of mine extraction (lithosphere peeling → waste accumulation → surface deformation) and environmental responses (geological hazards, pollution proliferation, and ecological degradation) and to quantify the relationship between the intensity of human activity and the degree of environmental impact.

Technology implementation:

- **Data sources:** These include geological exploration data (mineral distribution and rock structure), remote sensing image (multi-temporal satellite image comparison, such as Landsat 8 monitoring land cover change), UAV mapping (high-precision terrain relief data), and environmental monitoring (soil heavy metal content and groundwater heavy metal concentration).
- **Visual design:**
  Dynamic simulation of the mining process: 3D modeling shows the year-to-year expansion of a mine pit (overlay timeline, examples include "Mine pit depth increased from 50 m to 200 m in 2010–2020," and a description of damage to native landscapes caused by mining processes (e.g., blast sites, transport roads) (e.g., "Mountain slope increased from 30 m to 70 m, heightened risk of destabilization").
  Related map of environmental problems: This includes different layer superposition display "waste pile position → soil heavy metal excessive area (such as cadmium > 0.3mg/kg) → ground water pollution plume (nitrate concentration > 30mg/L)" and through the arrow connection "waste leaching → pollutants infiltration → ground water pollution" transmission path.
  Comparison of ecological impacts: Remote sensing images before and after mining (primary forest cover) are displayed side by side, with indicators of biodiversity (e.g., "Bird species decreased from 15 to 3 and vegetation cover decreased from 60 to 10%").

Typical case: This includes "green mine visualization monitoring platform" in a non-ferrous metal mining area in China. It integrates extraction progress, waste storage, soil remediation (e.g., vegetation restoration area), and environmental monitoring data (air quality and water quality) to help managers balance resource development with ecological protection through a dynamic dashboard of "extraction intensity — environmental impact — remediation effectiveness."

Scenario 3: Climate change and geological-environmental feedback loops — using permafrost melting as an example
Global warming causes accelerated thawing of permafrost at high latitudes and high altitudes. This geoscience process (the breakdown of the thermodynamic equilibrium) triggers an environmental chain reaction: Organic carbon in permafrost (which accounts for 50% of the global soil carbon stock) decomposes to release methane (CH4, the green house effect is 28 times greater than CO), land subsidence (infrastructure destruction), and wetland expansion (habitat change).

Objective of visualization: To reveal a positive feedback loop of climate change (temperature increase) → permafrost melting (geophysical process) → carbon release and environmental response (environmental science consequences) and to predict systemic risks under future scenarios.

Technology implementation:

- **Data sources:** These include climate models (CMIP6 forecasts global temperature changes over the next 50 years), permafrost monitoring networks (borehole thermometers record perennial frozen soil temperatures), remote sensing data (radar retrieval of frozen soil extent and thickness), and carbon cycle models (estimation of decomposition rates of organic carbon).
- **Visual design:**
  Dynamic map of permafrost melting: This uses color gradients (blue = stable permafrost, yellow = transition zone, and red = active melting zone), which shows spatial and temporal changes in the global distribution of the permafrost (e.g., "the rate of thawing of permafrost in the Arctic circle increased from 0.5 m/year in 1990 to 1.2 m/years in

2020"), superimposed a curve of increase in temperature (each 1°C increase in average annual temperature and 0.3 m increase in thawing depth).

Carbon release feedback cycle: The cycle of "thawing of permafrost → exposure of organic carbon → microbial decomposition → methane/ carbon dioxide release → enhancement of green house effect → further warming" is visualized by system dynamics model. Mark key points (e.g., "For every 1 ton of permafrost thawed, about 5 kg CH4 is released").

Infrastructure risk: Stack the locations of human works such as railways and pipelines to simulate deformation caused by frozen ground subsidence (e.g., "The Siberian gas pipeline costs an additional $200 million per year to maintain due to uneven ground subsidence").

Typical example: This includes the freezing soil module of the European Union's Copernicus Climate Change Service, which integrates satellite remote sensing and ground observation data. To demonstrate to policymakers the risk of releasing the equivalent of 50 billion tons of CO from permafrost in the event of a global warming of 2°C, and to promote coordinated strategies for mitigating warming and adapting to permafrost degradation.

10.4.3. *Design Principles and Technical Tools for Integrating Visualization*

Design principles:

- **A natural-humanitarian perspective:** This retains the objectivity of Earth science (e.g., the physical mechanisms of geological processes) while highlighting the human relevance of environmental science (e.g., industrial activities and policy interventions).
- **Dynamic coupled presentation:** This shows a cycle of "natural drives → human responses → environmental feedback" through a timeline animation or interactive slide (e.g., "mining activity → surface collapse → population relocation → ecological restoration").
- **Multiple source data fusion:** This is to unify data at different scales/ formats (e.g., to match macro-geomorphological changes from satellite

remote sensing with micro-pollution data from ground monitoring) to avoid information fragmentation.

Technical tools:

- **3D Earth Engine:** This includes Unity3D/Unreal Engine (building interactive global/regional models) and CessiumJS (lightweight 3D Earth visualization based on WebGL).
- **Coupled model interface:** This connects Earth science models (such as plate motion simulations and climate models) with environmental science models (e.g., pollution dispersion models and ecological models) through APIs to achieve real-time data interaction.
- **Spatiotemporal analysis tools:** This include ArcGIS Pro (multi-source spatial data overlay) and Python ecology (xarray for multi-dimensional climate data and matplotlib/seaborn for dynamic curves).

Visualization makes the Earth-environmental system more aware. Convergence of Earth and environmental science visualization is essentially a web of dynamic correlations between "hard constraints of nature" and "soft human intervention" — it makes visible how volcanic material pollutes the atmosphere. How soil is poisoned by the waste from mines can be sensed, and how methane from the thawing permafrost contributes to warming can be tracked. Such visualizations not only advance cross-cutting research in Earth system science (e.g., revealing the synergistic evolution of nature and humanities in the context of the Anthropocene) but also provide visual tools for "systems thinking" to address global challenges (e.g., climate change, resource depletion, and ecological crises). In the future, with the development of real-time monitoring technologies (such as IoT sensor networks) and digital twin, Earth-environment fusion visualization will further evolve toward "predicting-warning-decision support" and become a key support for the sustainable development of guardian lands.

10.5. Collaborative Visualization of Computer Science and Biomedicine: Digitally Driven Analysis of the Mysteries of Life

The intersection of computer science and biomedicine is one of the most disruptive directions in the field of science in the 21st century. From the explosion of gene sequencing data (the human genome contains 3 billion base pairs, and single-cell sequencing can resolve the heterogeneity of millions of cells) to the refinement of medical imaging (such as PET-CT to capture metabolism and anatomical structures at the same time), biomedical research is facing the challenge of "massive data but scarce insights." At the same time, algorithms (such as machine learning and graph computing), computing power (such as GPU parallel computing), and visualization technologies (such as 3D reconstruction and interactive interfaces) provided by computer science provide key tools for "dimensionality-correlation-prediction" of biomedical data. The core goal of this collaborative visualization is to transform the complex systems of biomedicine (such as molecular networks, organ function, and disease progression) into a "computable, interactive, and interpretable" digital picture, promoting the leap from "data-driven" to "knowledge discovery."

10.5.1. *Core Visualization Requirements and Challenges in the Collaboration Field*

The cross-visualization of biomedicine and computer science requires simultaneous processing of multimodal biomedical data (e.g., genomes, images, and electronic medical records) and high-dimensional computational model outputs (e.g., deep learning features and molecular dynamics simulations), and its challenges are mainly reflected in the following:

- **Data heterogeneity:** Genetic data is a discrete base sequence (A/T/C/G), imaging data is a continuous pixel matrix (such as T1/T2 weighted images of MRI), and electronic medical records are structured/unstructured text and values (such as blood pressure and medication records), which need to be uniformly characterized and fused.
- **Dimensional disaster:** Single-cell sequencing data can reach thousands of genes × millions of cells, and the voxel resolution of medical images is as high as microns (such as 0.1mm^3/voxel of brain MRI), and traditional two-dimensional charts cannot carry such high-dimensional information.
- **Dynamic complexity:** Biological processes (e.g., tumor growth, immune response) are time-dependent, and disease progression (e.g., cognitive decline in Alzheimer's disease) involves multi-stage characteristic changes, requiring dynamic tracking and association analysis.

Its core requirement is to combine the "computing power" of computer science with the "life logic" of biomedicine through visualization technology to reveal hidden patterns (such as disease subtype classification and drug target associations), mechanisms (such as how gene mutations affect protein function), and patterns (such as predictors of treatment response) hidden in the data.

10.5.2. *Typical Visualization Scenarios and Technical Strategies*

Scenario 1: Interactive exploration of genomic data – association from sequence to function

Background: The human genome contains about 20,000 protein-coding genes, but mutations in a single gene (such as pathogenic mutations in the BRCA1 gene) can cause cancer (such as breast cancer), while co-variants in multiple genes (such as TP53+KRAS combination) determine the degree of malignancy and treatment response of tumors. Traditional genetic data analysis relies on statistical tests (such as differential expression analysis), but it is difficult to visually present the whole chain association of "mutation-function-disease."

Visualization objective: To dynamically display the distribution characteristics of genomic variants (such as single nucleotide polymorphism, SNPs, and copy number variants, CNVs), correlate their corresponding protein functions (such as enzyme activity changes) with disease risk (such as cancer incidence), and assist researchers in identifying key pathogenic sites.

Technical implementation:

- **Data sources:** These include whole genome sequencing (WGS)/whole exome sequencing (WES) data (detection of SNPs, insertions, deletions, and other variations), public databases (such as TCGA tumor genome atlas, ClinVar disease-causing mutation library), and protein structure data (such as 3D structure predicted by AlphaFold).
- **Visual design:**
  - **Genome Browser:** This displays chromosomes (23 pairs) in a linear or ring graph, marks the mutation sites (such as red marks disease-causing mutations, and blue marks benign variants), and clicks to view the functional annotations of the corresponding genes (such as

"BRCA1: DNA damage repair-related"), protein structure (such as the three-dimensional position of the mutation site at the active site of the protein), and disease association (such as "the mutation accounts for 15% of breast cancer patients").

- **Multi-omics association network:** Nodes are used to represent genes, proteins, and diseases, and edges are used to represent interactions (such as "gene A regulates protein B" and "protein B participates in the pathway of disease C"), and nodes are dynamically adjusted through force-directed layout location, highlighting key modules (e.g., "mutation-enriched signal pathways").
- **Interactive filtering:** Users can filter and highlight relevant variant sites and associated pathways in real time by selecting mutation frequency (e.g., ">5%"), functional impact (e.g., "highly pathogenic"), or disease type (e.g., "Lung Cancer") via a slider.

Typical example: The National Center for Biotechnology Information (NCBI)'s "Genome Data Viewer" allows users to upload custom sequencing data, interactively compare reference genomes, and visualize individual-specific variants (e.g., "You have a missense mutation in your chromosome 17 TP53 gene that is associated with a poor prognosis in cancer databases").

Scenario 2: Intelligent analysis and 3D reconstruction of medical images: transformation from pixels to clinical decision-making

Background: Medical imaging (such as CT, MRI, and ultrasound) is the core tool for disease diagnosis, but traditional radiograph reading relies on the doctor's experience, making it difficult to quantify microlesions (such as 1mm grade nodules in early tumors) or dynamic changes (such as tumor volume reduction rate during treatment). Computer vision and deep learning technology can automatically detect lesions (such as lung nodules and cerebral hemorrhage), while 3D visualization can convert tomography images into three-dimensional models to assist in surgical planning and efficacy evaluation.

Visualization Goals: To automatically identify abnormal areas (e.g., tumors and infarct foci) in medical images, reconstruct their three-dimensional morphology and quantify key parameters (e.g., volume, density, and blood flow perfusion), and correlate patient clinical data (e.g., medical history and treatment plan) to support precision medicine decision-making.

Technical implementation:

- **Data sources:** These include multimodal medical images (such as CT tomography sequences, MRI T1/T2 weighted images, and PET functional metabolic maps), electronic medical records (such as tumor stage and previous treatment history), and clinical guidelines (such as NCCN Tumor Treatment Specifications).
- **Visual design:**
  - **Automatic detection and annotation of lesions:** This identifies suspicious areas in the image (such as lung nodules appear as ground glass density shadows in CT) through convolutional neural network (CNN) models (such as ResNet and U-Net), marks them with highlighted boxes or colors (such as red marking the probability of malignancy >80% of nodules), and outputs a confidence score.
  - **3D reconstruction and interoperation:** This converts the tomographic image sequence into a three-dimensional model (such as the 3D shape of the tumor) through volume rendering or surface rendering technology, supporting rotation, scaling, cutting (such as "peeling off skin and muscle, observing the adjacent relationship between tumor and blood vessel"), and superimposing measurement tools (such as volume calculation and maximum diameter measurement).
  - **Clinical correlation panel:** This displays key clinical information of patients (e.g., "tumor stage: T2N1M0" and "previous chemotherapy regimens: cisplatin + paclitaxel") next to the 3D model and shows the changes in imaging before and after treatment (e.g., "3 months after treatment, tumor volume decreased from 15 cm^3 to 6 cm^3, and density reduction suggests necrosis").

Typical case: Shanghai Zhongshan Hospital's "intelligent image-assisted diagnosis system" integrates lung CT AI detection (lung nodule detection rate >95%) and three-dimensional reconstruction function, doctors can observe the marginal characteristics of nodules (such as glitch signs suggesting malignancy), blood supply (enhanced scanning degree enhancement) through an interactive interface, and combine the patient's smoking history, family history, and other data to quickly formulate a needle biopsy or surgical plan, shortening the diagnosis time from the traditional 30 minutes to 5 minutes.

Scenario 3: Integrated visualization of biomedical multi-omics data — from molecular networks to disease mechanisms

Background: Biomedical research has entered the "multi-omics era" (genomics, transcriptomics, proteomics, and metabolomics), where single omics data (such as gene expression profiles) can only reflect a certain level of life activities, while multi-omics integration can reveal the cascade mechanism of "gene variation → transcriptional regulation → protein function → metabolite changes" (such as mutant genes upregulating specific signaling pathways in cancer, leading to metabolic reprogramming). Visualization of such data requires simultaneous processing of high-dimensional features (e.g., tens of thousands of genes + thousands of metabolites) with complex associations (e.g., "protein A is positively correlated with the concentration of metabolite B").

Visualization Objectives: To map multi-omics data into "molecular network diagrams" or "hierarchical association matrices" to highlight key nodes (such as pivot genes and core metabolites) and key pathways (such as disease-related signaling pathways) to assist researchers in discovering potential drug target dots or biomarkers.

Technical implementation:

- **Data sources:** These include multi-omics experimental data (e.g., gene expression of RNA-seq sequencing and metabolite concentration detected by LC-MS), public databases (e.g., KEGG pathway database and Reactome functional module), and clinical sample information (e.g., disease type and treatment response).
- **Visual design:**
  - **Molecular network diagram:** Nodes are used to represent genes, proteins, and metabolites, and edges are used to represent interactions (such as "gene transcription-regulated proteins" and "enzyme-catalyzed metabolite transformation"), and elements are distinguished by node size (expression/concentration) and color (omics type or functional category), highlighting key modules (e.g., "in breast cancer samples, key genes (ESR1, PIK3CA) in the estrogen receptor signaling pathway are significantly positively correlated with lipid metabolites (palmitic acid))."
  - **Hierarchical correlation matrix:** This arranges different omics data (such as gene expression matrix and metabolite concentration matrix)

by sample or functional module, shows the characteristic differences with heat map colors (red = high value, and blue = low value), and combines cluster analysis (such as hierarchical clustering) to group similar samples/modules (e.g., "samples with high expression of immune-related genes are clustered into one category, and their metabolites contain higher content of inflammatory factor precursors").

- **Dynamic path tracing:** Users can click on a specific node (e.g., "mutant gene TP53") to view its downstream influences transcripts (e.g., downstream gene CDKN1A expression downregulation), proteins (e.g., "p21 protein activity decreased"), and metabolites (e.g., "cell cycle-related metabolite R5P decreased") to form a complete logical chain of "mutant → function →phenotype."

Typical case: The Broad Institute's "Integrated Multi-Omic Viewer (IMOV)" supports researchers to upload genomic, transcriptome, and proteomic data from cancer patients, automatically match KEGG pathways, and generate interactive network maps, helping to discover that "in colorectal cancer, abnormal activation of WNT signaling pathway is closely related to fatty acid metabolism reprogramming, and the potential target is β-catenin protein." This key mechanism provides a theoretical basis for targeted therapy.

Scenario 4: Virtual patient and digital twin disease simulation — from personalized prediction to precision treatment

Background: The disease progression of each patient (e.g., risk of diabetes complications and tumor response to drugs) is affected by genetic background, environmental exposure, and lifestyle, and traditional "one-size-fits-all" treatment options are difficult to cover individual differences. Digital twin technology in computer science provides decision-making support for personalized medicine by building a "virtual patient model" (integrating genome, physiological parameters, and medical history data) to simulate disease progression and treatment response.

Visualization objectives: To dynamically display the changes in the physiological state (e.g., blood sugar level nad organ function), key disease indicators (e.g., tumor volume and immune cell infiltration) over time, and the simulation effect of different treatment regimens (e.g., drug dosage and timing of surgery) to assist doctors and patients in formulating optimal strategies.

Technical implementation:

- **Data sources:** These include multi-dimensional data of individual patients (such as gene sequencing results, continuous glucose monitoring CGM data, and medication records in electronic medical records), physiological models (such as blood circulation equations of the cardiovascular system and Gompertz model of tumor growth), and treatment guidelines (such as ADA diabetes management standards).
- **Visual design:**
    - **Virtual patient dashboard:** This shows historical changes and forecast trends of key physiological indicators (e.g., fasting glucose and glycated hemoglobin HbA1c) in the form of a timeline (e.g., "If the current diet is maintained, HbA1c will rise to 8.5% after 3 months") and risk scores (e.g., "Risk of progression of diabetic nephropathy: intermediate risk").
    - **Treatment simulation comparison:** This adjusts the treatment plan through interactive sliders (e.g., "Insulin dose increased from 10U to 15U" or "Surgery time is advanced from 2 months to 1 month after diagnosis") and displays simulation results in real time (e.g., "Blood sugar fluctuation is reduced by 30% after dose increase," "Early surgery can increase the 5-year survival rate from 60% to 75%").
    - **Organ-level visualization:** This is for specific diseases (such as liver cancer), 3D reconstruction of the liver model, and annotation of tumor location to simulate the effects of different treatments (such as radiofrequency ablation and targeted drug sorafenib) on tumor volume and surrounding tissues (such as blood vessels and bile ducts) (e.g., "80% of the tumor necrosis area after ablation, but mild narrowing of adjacent bile ducts narrow").

Typical case: The Mayo Clinic's "Diabetes Digital Twin System" has created a virtual model for patients that includes genetic risk (such as TCF7L2 gene mutations), daily diet (recorded through the app), and exercise volume (wearable device monitoring), predicts blood sugar fluctuations in real time, and recommends personalized interventions (such as "walking for 30 minutes tonight can reduce fasting blood sugar by 1.2 mmol/L tomorrow morning"), increasing the patient's blood sugar compliance rate from 45 to 72%.

10.5.3. *Design Principles and Technical Tools for Collaborative Visualization*

Design Principles:

- **Biomedical logic first:** Visualization should conform to the medical knowledge system (such as the causal chain of "gene → protein → function") and avoid purely technology-oriented "data stacking."
- **Dual adaptation between computing and clinic:** It not only provides researchers with high-dimensional data analysis tools (such as dimensionality reduction clustering and network topology analysis) but also designs an intuitive decision-making interface (such as risk score and treatment recommendation) for clinicians.
- **Privacy and security:** Biomedical data (especially genomes and medical records) involve patient privacy and need to be encrypted transmission, anonymization, and access rights control to ensure data security.

Technical Tools:

- **Data processing and analysis:** This includes Python ecosystem (Pandas processes table data, Scikit-learn implements machine learning, and Scanpy analyzes single-cell data) and R language (biostatistics and visualization packages such as ggplot2).
- **Visualization engine:** This includes Unity 3D/Unreal Engine (building 3D medical models and virtual scenes), D3.js/WebGL (interactive web visualization), and ParaView (large-scale rendering of scientific computing data).
- **AI integration:** This includes TensorFlow/PyTorch (deep learning model training) and SHAP/LIME (model interpretive tool to explain "why the AI thinks the nodule is malignant").

Visualization makes "computing-life" collaboration more efficient. The collaborative visualization of computer science and biomedicine is essentially a deep integration of "silicon-based computational intelligence" and "carbon-based life mysteries" — it transforms the base sequence of the genome into an interactive functional map, upscales the pixel matrix of medical images into an operable three-dimensional model, condenses the massive information of multi-omics data into a traceable mechanism network, and makes the digital twin of virtual patients a "rehearsal sand table" for precision medicine. This kind of visualization not only accelerates breakthroughs in basic research (such as new target discovery and disease mechanism analysis) but also promotes changes in clinical practice (such as personalized treatment and early risk warning) and finally achieves the ultimate goal of "protecting life and health with the power of computing." In the future, with the integration of quantum computing, edge intelligence, and biosensing technology, biomedical collaborative visualization will further evolve in the direction of "real-time, dynamic, and individualized" and become the core infrastructure of the digital health era.

10.6. Extended Visualization of Social and Human Sciences: Digital Decoding from Urban Pulse to Cultural Memory

Social sciences and humanities focus on the "dynamic evolution of human behavior, social structure, and cultural significance," and their research objects are often characterized by strong subjectivity (such as cultural values), high heterogeneity (such as behavior patterns of different groups), and complex spatiotemporal relationships (such as social stratification of urban space and historical accumulation of cultural heritage). Traditional research relies on textual descriptions, questionnaires, or qualitative analysis, making it difficult to intuitively present "the laws of group behavior," "deep associations of cultural symbols," or "immersive experiences of historical scenes." By combining social behavior data (such as mobile phone signaling and social media check-in), humanistic elements (such as three-dimensional models of ancient buildings and document semantic networks), and spatial and temporal dimensions, visualization technology has become a key bridge connecting "data science" and "humanistic care."

10.6.1. *Core Visualization Needs and Challenges in the Extension Field*

The intersection visualization of social sciences and humanities needs to deal with the dynamics of human activities (such as instantaneous changes in crowd flow), the metaphorical nature of cultural significance (such as the symbolic function of ancient buildings), and the heterogeneity of multi-source data (such as the subjective feedback of mobile phone signaling data and questionnaires), and its challenges mainly include the following:

Privacy and ethics of behavioral data: Urban population movement data (such as GPS tracks and base station positioning) involve personal privacy, and macro patterns need to be extracted under the premise of protecting sensitive information.

Abstract transformation of cultural semantics: The value of cultural heritage (such as "the architectural layout of the Forbidden City reflects the feudal etiquette") needs to be visualized to transform the implicit cultural logic into intuitive visual language.

Nested coupling at the spatiotemporal scale: Social phenomena (such as peak traffic in business districts during holidays) are driven by both short-term events (such as promotional activities) and long-term historical context (such as the centennial evolution of urban functional areas).

Its core goal is to transform the "group law" of social sciences and the "cultural connotation" of humanities into "observable, comparable, and reflective" digital pictures through visualization technology, assist researchers in discovering the mechanism of social operation, and promote public empathy and understanding of culture.

10.6.2. *Typical Visualization Scenarios and Technical Strategies*

Scenario 1: Analysis of spatial–temporal patterns of urban crowd flow — decoding the "digital footprint" of urban life

Background: Cities are the most complex systems with intensive human activities, and the flow of people (such as commuting, leisure, and consumption) not only reflects economic vitality and social structure but also affects transportation planning, public service allocation, and the risk of epidemic transmission. Traditional research obtains local data through questionnaire surveys or fixed-point counting, which makes it difficult to capture the dynamic law of the whole city. New data sources such as mobile phone signaling data (recording base station switching tracks), subway card swiping records, and shared bicycle tracks provide the possibility for large-scale crowd flow analysis.

Visualization objectives: To dynamically display the spatial–temporal distribution characteristics of urban crowd flow (such as weekdays/weekend differences and holiday hot spot transfers), identify key flow patterns (such as "separation of work and housing" and "business district agglomeration"), and assist urban planning and management decision-making.

Technical implementation:

- **Data sources:** These include anonymized mobile phone signaling data (recording the user's switching time and location between different base stations, and reflecting coarse-grained movement trajectories), subway/bus IC card swiping records (accurate to stations and time stamps), shared bicycle GPS tracks (micro-scale short-distance travel paths), and POI (points of interest) data (marking shopping malls, office buildings, parks, and other functional areas).
- **Visual design:**
  - **Spatial–temporal heat map:** This shows the crowd gathering areas in different time periods (such as the commuting peak from 8:00 to 9:00 on weekdays, and the business district from 15:00 to 17:00 on weekends) with a color gradient (red = high density, blue = low density) and superimposes the boundaries of urban functional areas (such as "red high-density areas are concentrated in the office building and Guomao Shopping Center on Financial Street").
  - **Flow trajectory network:** The base station/station is used as the node, and the crowd movement path is used as the edge, and the main flow direction is displayed through dynamic line animation (e.g., "the crowd flows from Changping residential area to Haidian Science and Technology Park in the morning peak, and the reverse flow during the evening peak"), and the width is used to express the traffic size (the thicker the line, the higher the flow).
  - **Comparison panel:** This displays different dates (such as jobs, side by side day vs. weekend), changes in crowd flow patterns in different seasons (such as extended summer nightlife) or emergencies (such as during epidemic lockdowns), and marks key differences (e.g., "After 20:00 on weekend nights, the crowd density in Sanlitun is 3 times that of weekdays").

Typical case: The "Crowd Flow Digital Twin Platform" of the Beijing Institute of Urban Planning and Design integrates the anonymous mobile phone signaling data of tens of millions of users in the city and simulates the urban "vitality map" of the city on weekdays and weekends in real time — managers can view "which subway stations exceed the threshold in the morning peak rush hour (additional trains need to be dispatched)" and "which business districts gather at night but lack lighting (need to strengthen security)" and adjust bus routes or open temporary parking lots based on the prediction model.

Scenario 2: Immersive digital display of cultural heritage — an interactive experience that makes history "come alive"

Research background: Cultural heritage (such as ancient buildings, ruins, and cultural relics) carries the historical memory and cultural identity of human civilization, but traditional display methods (such as static display in museum glass cabinets and text explanation boards) are difficult to convey the construction skills, social functions, and aesthetic values behind them. The combination of digital technologies (e.g., 3D scanning and VR/AR) and immersive visualization (e.g., light and shadow simulation and multi-sensory interaction) can transform static heritage into dynamic "accessible, operable, and dialogueable" cultural fields.

Visualization goal: Through high-precision three-dimensional reconstruction and multimodal interaction, the original historical appearance of cultural heritage (such as the structural details of Tang Dynasty buildings), usage scenarios (such as the court meeting ceremony of the Taihe Hall of the Forbidden City) or production process (such as the fan casting method of bronzes) is restored to assist the public in understanding its cultural connotation and provide a scientific basis for protection and restoration.

Technical implementation:

- **Data sources:** These include LiDAR scanning (obtaining millimeter-level 3D point cloud data of ancient buildings/sites), multispectral photography (recording the color layering and disease traces of murals/cultural relics), historical documents (such as the architectural regulations recorded in "Building the French Style"), and oral history interviews (the process described by the inheritors).
- **Visual design:**
  - **High-fidelity 3D reconstruction:** This builds a complete digital model of the ancient building based on point cloud data (e.g., each bucket arch member of the Yingxian Wooden Pagoda can be

dismantled separately), superimposes the material texture (restore the sense of wood over time through high-definition image maps), and marks the key structures (such as "the seismic principle of mortise and tenon nodes" and "the direction of the dark ditch of the drainage system").

- **Immersive scene simulation:** Through VR headsets or CAVE projection systems, users can "immerse" themselves in historical scenes — such as "standing on the Hongqiao in Bianjing in the Song Dynasty, observing the shops and pedestrians on both sides of the Bian River" and "participating in the winter solstice worship ceremony in the Forbidden City in the Ming Dynasty, and feeling the hierarchical order in the spatial layout" — or combine ambient sound effects (such as bells and market noise) and dynamic light and shadow (such as the projection of sunlight through the window lattice at sunrise) to enhance the sense of presence.
- **Interactive cultural interpretation:** Users can trigger the "information layer" through gesture/voice interaction — click on a component of an ancient building (such as a bucket arch) and pop up an animation to demonstrate its construction steps ("erect the column network first, then install the timber, and finally splice the bucket arch"), and scan the decorations on the surface of the cultural relics (such as the gluttonous pattern on the bronze) to show their symbolic meaning ("Taotie is the totem of power and divine power in the Shang and Zhou dynasties").

Typical case: The "Digital Dunhuang" project of the Dunhuang Research Institute transforms 30 classic caves of the Mogao Grottoes (including the special caves that are prohibited from opening) into digital copies through high-precision 3D scanning and multispectral imaging technology, and users around the world can "roam" the caves through the web page or VR equipment, observe the flying clothes streamers (resolution up to 0.1mm) in the mural, the painting level of the Buddha statues (the original color before fading), and view the interpretation videos of scholars on the themes of the murals (such as Buddhist stories and secular life). Let millennium art break through the limitations of geography and time and "live in the present."

Scenario 3: Dynamic visualization of social network relationships — revealing the implicit structure of group interaction

Background: In social science research, “relationships” are the core variables (e.g., interpersonal trust, organizational collaboration, and cultural communication), but traditional analysis relies on questionnaires (highly subjective) or small-scale observations (limited sample). Social media data (such as Weibo following/forwarding and WeChat group chat records) and mobile phone call network (base station positioning user communication frequency) provide an objective basis for large-scale social network analysis. Visualization technology can transform abstract “relationship networks” into intuitive “node-edge” graphs, revealing hidden community structures and influence transmission paths.

Visualization goal: To dynamically display the connection relationship (edge) of individuals (nodes) in the social network and the aggregation characteristics of groups (such as community circles and opinion leaders) and analyze the paths and mechanisms of information dissemination, cultural dissemination, or crisis diffusion.

Technical implementation:

- **Data sources:** These include social media APIs (such as Weibo users’ watchlists and retweet comment records), mobile operator data (anonymized call/SMS frequency and duration), and offline event sign-in data (such as exhibition participants’ interaction records).
- **Visual design:**
  - **Network relationship graph:** This uses nodes to represent individuals (such as Weibo bloggers and community residents), edge to represent relationship strength (such as “mutual attention” as a thick red line and “one-way follow” as a thin blue line), and automatically cluster similar groups (such as “celebrity fan group” and “academic research circle”) through force-oriented layout. The node size indicates influence (such as the number of fans and retweets), and the color distinguishes the community category (e.g., red = interest community, and blue = professional community).
  - **Dynamic communication simulation:** This uses “information/cultural elements” (such as a hot search topic or a traditional custom) as “seeds,” animated to demonstrate its diffusion process in the network (e.g., “posted from 1 blogger → retweeted by 100 fans → spread to related communities”), and marked key nodes (e.g., “Opinion leader’s retweet increases the spread speed by 3 times”).

- **Time slice comparison:** This uses the timeline slider to observe the dynamic changes in network structure (e.g., "During the epidemic, the density of community residents' mutual aid network has increased significantly, while cross-community connections have decreased") or compares the differences in communication patterns of different events (e.g., "celebrity scandal" and "public welfare initiative").

Typical case: In the "community vitality network analysis" of an urban planning project, researchers used mobile phone signaling data (recording residents' base station stops on weekdays and weekends) to construct a "social-space" correlation network — it was found that although the residents of some old communities were physically close, their daily activities were isolated (such as only going back and forth between home and the vegetable market), while the resident network around the new complex showed multi-community intersection (such as frequent interaction between office workers, housewives, and tourists), which was the goal of "promoting community integration through public space design" data support.

Scenario 4: Digital Restoration and Contrast of Historical Scenes — "Scene Reproduction" Across Time and Space

Research background: Historical and archaeological research often faces the limitations of "site damage" (such as the ancient city wall only retains the foundation site) or "documentary abstraction" (such as the textual description of the urban layout in ancient books), and it is difficult to intuitively present the real scene of the historical period. Through the fusion of multi-source data (such as historical maps, archaeological excavation reports, and restoration maps of ancient paintings) and 3D modeling technology, the street style, architectural shape, or life scene of the historical city can be reconstructed, assisting scholars to verify hypotheses, and at the same time allowing the public to "travel" to the past to feel the cultural context.

Visualization goal: To transform historical documents and archaeological data into high-precision three-dimensional scenes (such as the market pattern of Chang'an City in the Tang Dynasty, and shops and pedestrians in the Song Dynasty market) and reveal the continuity and fracture of urban evolution through superposition and comparison with modern cities.

Technical implementation:

- **Data sources:** These include historical maps (such as vectorized data of the "Complete Map of Qianlong Capital"), archaeological excavation reports (the hierarchical relationship of the site and the distribution of excavated cultural relics), ancient paintings/murals (such as the urban life of the Northern Song Dynasty depicted in the "Qingming River Map"), and modern surveying and mapping data (satellite images and topographic maps of contemporary cities);
- **Visual design:**
  - **3D reconstruction of historical scenes:** This restores the main structure of the building based on archaeological data (such as the rammed earth platform foundation of the Tang Dynasty palace and the wooden eaves of the Song Dynasty shops), adds decorative elements (such as lanterns and signboards) with reference to the details of ancient paintings, and enhances the sense of realism through material simulation (such as the rough texture of rammed earth and the blue-gray luster of tiles).
  - **Superposition and comparison of ancient and modern times:** This includes superimposing historical scene models with modern satellite images (e.g., "superimposing the Suzaku Street of Chang'an City in the Tang Dynasty on the road network within today's Xi'an city wall"), marking the changed areas (e.g., "the former imperial city area is now a modern government office area"), or demonstrating the process of urban expansion through timeline animations (such as "from the inner city of Beijing in the Ming and Qing dynasties to the outer city of the Republic of China, and then to the expansion of the Five Rings in modern times").
  - **Situational interactive experience:** Users can "walk" in historical scenes (such as walking along the wharf along the Bian River in the Song Dynasty), observe the professional activities at that time (such as boatmen loading and unloading goods, cargo merchants selling), and click on specific buildings (such as restaurants) to view the business model recorded in the literature ("the market opens every afternoon, focusing on north-south and north-south cuisine").

Typical case: Xi'an's "Digital Tang Dynasty Never Sleeps City" project, through three-dimensional scanning and historical research, restores the layout of the market around the Wild Goose Pagoda in the Tang Dynasty (including the types of shops in the East City and the West City, and the gathering area of Hu merchants), and tourists can "see" the virtual Tang Dynasty caravan and dancer performances in modern scenic spots through AR glasses, and even participate in the "virtual imperial examination" interactive game, so that history and culture can enter the real experience from books.

10.6.3. *Design Principles and Technical Tools for Extended Visualization*

Design Principles:

- **Humanistic care first:** Visualization needs to respect cultural sensitivity (such as avoiding misreading the cultural heritage of ethnic minorities) and enhance the emotional resonance of the public through "storytelling narrative" (such as "the story of a craftsman family and the construction of the Forbidden City").
- **Data-interpretation collaboration:** Social behavior data (such as crowd flow) should be combined with sociological theories (such as the "central location theory" to explain the distribution of business districts), and cultural heritage data (such as ancient buildings) should be related to historical documents (such as the construction regulations in the "Kao Gong Ji") to avoid "having pictures without meaning."
- **Privacy protection and ethical compliance:** Population movement data need to be anonymized (such as aggregated to the community level rather than individual tracks), and cultural heritage data need to comply with cultural relics protection regulations (such as not displaying undisclosed tomb details).

Technical Tools:

- **Social network analysis:** This includes Gephi (network graph visualization), NetworkX (Python library for complex network computing).
- **3D reconstruction and immersive display:** This includes Reality Capture (high-precision 3D scan modeling), Unity 3D/Unreal Engine (VR/AR interactive development), and SketchUp (rapid prototyping).
- **Spatiotemporal data processing:** This includes PostGIS (geospatial database) and Python ecosystem (Pandas processes time series, and Folium generates interactive maps).

Visualization makes "social-humanistic" research warmer. The extension visualization of social sciences and humanities is essentially a deep integration of "digital rationality" and "humanistic temperature" — it makes the flow trajectory of urban people a clue to understand the social structure, allows the ruins of cultural heritage to regain its former glory, allows the relationship map of social networks to reveal the mystery of group interaction, and allows the restoration of historical scenes to cross the gap of time and space. This kind of visualization not only provides scholars with tools to "go from data to insight" but also builds a bridge for the public "from observer to participant" and finally achieves the ultimate goal of "protecting the diversity of human civilization with digital technology." In the future, with the development of the Metaverse, generative AI (such as large language models interpreting ancient documents), and brain–computer interfaces (simulating the emotional experience of historical scenes), the social-humanistic extension visualization will further evolve in the direction of "immersive, intelligent, and empathetic," becoming a digital link connecting the past, present, and future.

Part 5

Frontiers and Prospects: The Future of Scientific Visualization Driven by Digital Media

Chapter 11

Frontier Technology Trends: Scientific Visualization Toward a New Era of Intelligence and Integration

11.1. Introduction: When Traditional Visualization Meets the Intelligent Revolution — How the "Visual Tools" of Scientific Exploration Evolved into "Cognitive Partners"

The history of scientific visualization is a history of breaking through the boundaries of cognition and expanding the boundaries of visual expression. From static "data presentation tools" to interactive "exploration and analysis platforms," visualization technologies have consistently resonated with the evolution of scientific research paradigms. However, as scientific research has entered the era of "data-intensive scientific discovery" (the fourth paradigm), the size of scientific data has increased exponentially (e.g., the LHC produces petabytes of collision data per second), and the dimensions have become increasingly complex (multi-physics field coupling). With multi-omics data convergence and increasing dynamics (e.g., time-series evolution modeled for climate change and real-time dynamics of biological cells), traditional visualization technologies are facing unprecedented challenges — how to extract critical information quickly from massive amounts of data? How to visualize nonlinear correlations between high-dimensional variables? How can we support more natural and in-depth interaction between researchers and visualization systems? These issues are driving the evolution of scientific visualization toward more intelligent and integrated technologies (Zhang *et al.*, 2021; Chen & Müller, 2022).

Today, we are at a new starting point for scientific visualization: artificial intelligence, virtual reality (VR/AR). The integration of digital twinning and Internet of things is upgrading the traditional "static/interactive visualization tools" to "intelligent visualization system with cognitive ability," promoting scientific visualization to a new era of "intelligence and integration." In this new era, visualization is no longer just a "presenter" of data but also a "collaborator" of scientific discovery, the "interpreter" of complex systems as well as the "responder" to user needs — the ability to automatically identify hidden patterns in data (such as disease-related features in gene expression data) through machine learning. VR/AR technology to build immersive three-dimensional exploration space (such as the microscopic world of molecular structure or the geological structure of the Earth), relying on digital twinning to achieve real-time synchronization of physical systems and virtual models (such as visualization of the state in service, and failure prediction of plant equipment), IoT sensing dynamically collects multi-source heterogeneous data (such as environmental monitoring and traffic flow data in smart cities), ultimately providing researchers, engineers, and the public with a more precise, intuitive, and immersive scientific cognitive experience.

This evolutionary trend is reflected in three dimensions of technological breakthrough and fusion innovation:

First is intelligence: a significant shift from "human-driven" to "AI-empowered" capabilities. Artificial intelligence technologies, particularly deep learning and generative AI, are reshaping the core processes of scientific visualization — particularly during the data preprocessing stage. AI algorithms can automatically detect outliers, fill in missing data, and extract key features (such as separating effective information from noisy sensor signals). In the visual design phase, intelligent recommendation systems based on user goals and data characteristics can automatically generate the optimal visual coding scheme (such as automatic selection of color mapping, graphic type, and interaction logic). During the exploration and analysis phase, an AI-driven visualization system can facilitate natural language interactions (such as a user asking, "Which genes are most closely associated with cancer metastasis?"), quickly generate a corresponding visual diagram, or even proactively push potential scientific hypotheses (e.g., "the conformational changes in this protein may be related to drug resistance"). For example, in the field of biomedicine, AI-assisted visualization tools have been able to automatically identify areas of diseased cells within microscopic images and overlay the results of pathological analysis. In climate science, generative AI can simulate the global temperature change curve under different greenhouse gas emission scenarios, assisting researchers in predicting future trends (Kumar *et al.*, 2022; Hansen *et al.*, 2023).

Second is convergence: experience upgrades from "single technology" to "multimodal synergy." The convergence of cutting-edge technologies is breaking the boundaries of traditional visualization to create more immersive and interactive scientific exploration environments — virtual reality (VR) and augmented reality (AR) technologies are "embedding" users in 3D data fields, such as inside a molecule or a geological structure. This enables intuitive micro exploration through gesture manipulation and natural line tracking. Digital twin technologies build virtual mirrors of physical systems (such as urban transportation networks or human organ models) and enable real-time data synchronization and dynamic predictions (such as simulating evacuation schemes for traffic congestion or the evolution of organ lesions). The Internet of Things (IoT) dynamically collects multi-source heterogeneous data (such as environmental parameters, device state, and human behavior) through the sensor network to provide real-time updated information source for visualization. Mixed reality (MR) further integrates virtual and real elements (such as superimposing a three-dimensional anatomical model of the patient in a real surgical scene) to assist doctors in accurate operations. For example, in the aerospace field, engineers can "enter" the internal structure of the rocket engine through VR equipment to observe the airflow and temperature distribution of the combustion chamber. In smart city management, decision makers can view real-time air quality index and human flow density thermal maps superimposed on real streets through AR glasses.

Third is universalization: an extension of the value from "professional tools" to "accessible to all." With the intelligent and integrated development of visualization technology, its application scene is expanding from the professional scientific research and engineering field to the broader public domain — scientific visualization is no longer limited to laboratories or data centers but through mobile devices, web platforms, and immersive experience museum to the Volkswagen, become the popularization of scientific knowledge, improve the scientific literacy of the public an important carrier. For example, popular science visualizations for teenagers make complex cosmological knowledge interesting through gamified design and simple interactions (like clicking on a planet to see how it formed); public-facing environmental visualization tools (such as real-time air quality maps and carbon footprint calculators) help people visualize the impact of their actions on the planet; personalized health visualization for medical patients (such as checking their electrocardiogram dynamics and disease risk trends through mobile app) allows each person to become a "manager" of their own health. This universal tendency not only expands the influence of scientific visualization but also promotes the deep integration of science and society so that the results of scientific exploration truly benefit everyone (Liu *et al.*, 2023; NASA, 2022).

This chapter focuses on the latest technological trends in scientific visualization and systematically examines how these innovations are driving visualization toward a new era of intelligence and convergence. Interaction and analysis reveal how VR/AR and digital twin technologies reconstruct the immersive experience of scientific exploration and analyze how the Internet of Things and edge computing support real-time dynamic visualization. At the same time, attention should be paid to the ethical challenges arising from the integration of these technologies (such as data privacy protection and the interpretability of AI-driven decisions) as well as their future development directions (such as universal scientific visualization assistants and interdisciplinary collaborative intelligent platforms). Through these discussions, the reader will see how scientific visualization has evolved from "assistive device" to "cognitive partner," from "professional field" to "social panorama," and eventually become an important engine of scientific advancement and human civilization (Zhao *et al.*, 2023; UNESCO, 2022).

In this new era of intelligence and integration, the mission of scientific visualization is not only "let data be seen" but also "let knowledge be understood," "let innovation be inspired," and "let science be shared." More immersive interactive experiences and more universal value delivery are writing a new chapter in human exploration of nature and ourselves (Zhang, *et al.*, 2023b; ESA, 2022).

11.2. Artificial Intelligence Empowerment: A Paradigm Shift from "Assisted Display" to "Intelligent Insight"

The rapid development of artificial intelligence (AI) technology is profoundly changing the logic of scientific visualization — traditional visualization takes the one-way mapping of "data → graphics" as the core, while AI-driven intelligent visualization realizes automation and intelligence from data preprocessing, pattern discovery to visual generation through the two-way interaction of "data → cognition," and has become an "intelligent assistant" for researchers and engineers.

11.2.1. *Intelligent Data Preprocessing and Feature Extraction*

Traditional preprocessing relies on manual experience (such as manual selecting filter parameters and setting rules for filling missing values), while AI technology can optimize the process through automated algorithms. For example:

- **Intelligent dimensionality reduction and feature selection:** The deep learning-based autoencoder can automatically learn low-dimensional representations of high-dimensional scientific data (such as single-cell RNA sequencing data and meteorological multivariate timing data), retaining key information while removing redundant noise. The Attention Mechanism automatically identifies the characteristic variables that contribute the most to scientific objectives (such as disease classification and climate prediction) and assists in visualization to focus on the core dimensions (Jordan & Mitchell, 2021; Goodfellow *et al.*, 2022).
- **Dynamic data repair and interpolation:** For missing data in sensor networks (such as intermittent data in marine buoy monitoring), the spatio-temporal distribution law of the data can be learned by generating adversarial networks (GANs) to generate reasonable filling values that meet physical constraints. Recurrent neural networks (RNNs) or transformer models can predict the values of missing time points based on historical time series data, ensuring the continuity of dynamic visualization (Goodfellow *et al.*, 2022; Hochreiter & Schmidhuber, 2021).

11.2.2. *Automated Visualization Generation and Optimization*

AI can automatically recommend or generate optimal visualizations based on data characteristics and user goals, lowering the operational threshold for non-professional users. Typical techniques include the following:

- **Visual recommendation system:** Analyze data types (e.g., numeric, sub-type, and spatial-temporal type), scientific objectives (e.g., trend analysis, correlation discovery, and distribution display), and user preferences (e.g., simplicity and interactivity) through machine learning models (e.g., classifiers and regression models), automatically recommend the most suitable chart types from a library of predefined visualization templates (e.g., line charts, heat maps, and 3D scatter plots), and even generate customized visualization layouts (e.g., multi-view linkage schemes). For example, business tools such as Tableau have integrated this feature, and after users upload data, the system automatically suggests "showing the percentage of sales in different regions in a stacked bar chart (Mackinlay *et al.*, 2020; Heer & Shneiderman, 2021)."
- **Automatic Parameter Optimization:** In complex visualization scenarios (e.g., streamline density in flow field visualization and color mapping range in molecular dynamics simulation), AI algorithms (e.g., reinforcement learning and Bayesian optimization) can automatically adjust key parameters, making the visualization results clearer and more informative. For example, the reinforcement learning model dynamically optimizes the distribution and density of the starting point of the streamline, which not only avoids the visual confusion caused by the overlap of the streamline but also fully presents the core area of the vortex when showing the turbulent structure (Silver *et al.*, 2021; Sutton & Barto, 2022).

11.2.3. *Intelligence Analysis and Knowledge Discovery*

AI can not only "display data" but also "understand data" — through deep learning models to mine potential patterns and scientific laws in the data, assisting users in discovering new knowledge. For example:

- **Anomaly detection and early warning:** In scenarios such as industrial equipment monitoring (such as engine vibration signals) and environmental monitoring (such as air quality sensor networks), anomaly detection models based on convolutional neural networks (CNNs) or long short-term memory networks (LSTMs) can automatically identify abnormal data points that deviate from normal patterns (such as vibration anomalies before equipment failure, PM2.5 spikes in pollution emergencies) and help users quickly locate problems by visualizing highlighted anomalous areas (such as red marked anomalous time points or spatial locations) (Esteva *et al.*, 2021; LeCun *et al.*, 2022).

- **Correlation rule mining:** In biomedicine (e.g., gene-disease association) and geoscience (e.g., climate factor-ecological response coupling), correlation rule learning algorithms (e.g., an improved version of Apriori's algorithm) can automatically discover hidden associations between variables (such as "a specific gene combination is significantly correlated with the incidence of a certain cancer" and "the probability relationship between sea surface temperature anomalies and typhoon intensity enhancement") and visually present the strength and direction of associations through network diagrams, heat maps, and other visual forms (Han *et al.*, 2020; Getoor & Taskar, 2007).

11.3. Immersive and Interactive Experiences: Sensory Upgrades from "Flat Observation" to "Deep Engagement"

With the development of virtual reality (VR), augmented reality (AR), mixed reality (MR), and multimodal interaction technology, scientific visualization is evolving from the traditional "two-dimensional display on screen" to "immersive three-dimensional space experience."

11.3.1. *In-Depth Application of Virtual Reality (VR) and Augmented Reality (AR)*

- **VR immersive exploration:** In the microscopic field (such as protein molecular structure and atomic-level material defects), users can freely rotate and scale molecular models in three-dimensional space through VR headsets and controllers to observe the dynamic processes of electron cloud distribution and chemical bond formation. In the macroscopic field (such as galaxy evolution and geological tectonic movement), VR technology can build a "digital universe" or "virtual earth," and users can "shuttle" between different galaxies or go deep into the earth's crust to observe the activities of plate boundaries. For example, the "Mars VR Experience" project developed by NASA allows users to walk on the surface of Mars from the perspective of an astronaut, combining real terrain data and scientific exploration results to intuitively feel the geological characteristics and potential habitable environment of Mars.

- **AR augmented reality assistance:** In engineering practice (such as mechanical equipment maintenance and construction sites) and science education (such as anatomy teaching and geological specimen observation), AR technology superimposes virtual visual information (such as internal structure of equipment and 3D restoration models of fossils) onto real objects through tablets or smart glasses, assisting users to obtain key information in real time. For example, maintenance engineers can scan the faulty engine through AR glasses and see the 3D disassembly diagram and fault point annotations superimposed on real parts. When medical students observe human skeletal specimens through AR devices, they can click on any bone to view its name, function, and association of adjacent tissues.

11.3.2. *Multimodal Interaction and Nature Perception*

While traditional interactions rely on discrete devices such as mice and keyboards, future scientific visualization will support more natural multimodal interactions, including the following:

- **Gesture and somatosensory interaction:** This captures the user's gesture movements (such as grabbing, rotating, and dragging) through the depth camera (e.g., Kinect and Leap Motion) to achieve direct control of 3D visualization objects. Somatosensory technologies (such as Microsoft Kinect) can also recognize the user's body posture (e.g., leaning forward to see details, stepping back to see the big picture) and dynamically adjust the visual perspective and level of detail (LOD).
- **Voice and eye interaction:** Voice commands (such as "enlarge this area" and "display temperature distribution") can quickly trigger visual operations, reducing the user's operational burden. Eye tracking

technology (which monitors the focus of the user's gaze through an infrared camera) automatically highlights areas of concern (such as areas of cells that researchers have been looking at for a long time) and adjusts the priority of information presentation (such as overlaying detailed parameter labels on that area).

- **Haptic feedback and force feedback:** In medical surgery simulation (such as virtual cardiac intervention), geological disaster drill (such as seismic fault simulation and tactile sensation), and other scenarios, the hardness, friction, and vibration of objects are simulated through force feedback devices (such as tactile gloves and vibration platforms), allowing users to "touch" virtual scientific objects (such as feeling the softness of tumor tissue and the impact of seismic waves), enhancing the sense of immersion and realism of operation.

11.3.3. *Social and Collaborative Visualization*

Scientific exploration often requires team collaboration, and future visualization systems will support real-time sharing and interaction between multiple people. For example, scientific research team members can enter the same virtual laboratory at the same time through the cloud platform, operate different visual views (such as one person viewing molecular dynamics simulation results, and the other person analyzing spectral data), and collaborate on analysis through real-time voice communication and annotation sharing (such as labeling key atoms or abnormal regions on a common 3D model). In the public science popularization scene, visitors can enter the immersive exhibition through social media links to explore scientific phenomena with other online users (such as operating typhoon path simulations together and discussing the impact of climate change), forming a new model of "scientific socialization."

11.4. Cloud Computing and Edge Computing: The Computing Power Revolution Supporting Large-Scale Visualization

The explosive growth of scientific data (such as high-resolution satellite remote sensing images, large-scale particle physics experiment data, and real-time sensor network streaming data) has put forward higher requirements for visualization computing power — traditional local computing is difficult to handle the real-time rendering and interaction of terabytes or even petabytes of data. The coordinated development of cloud computing and edge computing provides flexible and efficient computing power support for large-scale scientific visualization.

11.4.1. *Cloud Computing: Elastic Resources and Distributed Processing*

Cloud computing platforms (such as Alibaba Cloud, AWS, and Google Cloud) provide nearly unlimited storage and computing resources, supporting the elastic expansion of scientific visualization systems. Its core benefits include the following:

- **Large-scale data storage and management:** Distributed file systems (such as HDFS) and object storage (such as OSS) are used to store massive scientific raw data (such as multi-temporal satellite imagery and long-term meteorological observation records), and unified management of structured and unstructured data are achieved through data lake technology.
- **Distributed Computing and Rendering:** This utilizes cluster computing (e.g., Spark and Hadoop) to process complex data preprocessing tasks (e.g., multivariate statistical analysis, grid data interpolation) in parallel, and accelerate real-time rendering of 3D models (e.g., complex flow field visualization, dynamic display of molecular dynamics simulations) through GPU clusters (e.g., NVIDIA A100 GPU cloud service). Based on WebGL or WebGPU cloud rendering technology, the rendering results are transmitted to user terminals (such as browsers and mobile devices) in the form of video streams, lowering the threshold for local hardware (users can view high-precision visualization without high-end graphics cards).
- **On-demand services and cost optimization:** Scientific research institutions and small and medium-sized enterprises can flexibly rent cloud computing resources according to project needs (such as temporarily increasing GPU computing power to cope with emergency simulation tasks), avoiding huge investment in building their own high-performance computing centers and maximizing cost-effectiveness.

11.4.2. *Edge Computing: Low Latency and Real-Time Response*

For scientific scenarios with extremely high real-time requirements (such as online monitoring of industrial equipment, environmental perception in autonomous driving, and earthquake early warning systems), edge computing realizes local preprocessing and rapid visualization of data by deploying lightweight computing units near data sources (such as factory sensor nodes, vehicle terminals, and earthquake monitoring stations). Typical applications include the following:

- **Real-time data filtering and compression:** Edge nodes perform real-time cleaning (such as noise removal and outlier removal), down sampling (reducing data frequency to reduce transmission), and compression (such as using lossy or lossless compression algorithms) to transmit only critical information (such as abnormal data points and trend summaries) to the cloud or local servers, reducing network bandwidth pressure and improving response speed.
- **Local visualization and quick decision-making:** In industrial control centers, edge computing devices can directly generate real-time visualization interfaces for key parameters (such as dynamic dashboards of equipment temperature and pressure) to assist operators in real-time judgment of equipment status. In the disaster warning scenario, the edge node quickly generates a simplified visual warning map (such as high-risk area labeling) based on local sensor data (such as the time of initial arrival of seismic waves and rainfall threshold) and sends warnings to relevant personnel through local displays or mobile terminals to buy valuable time for emergency response.

11.4.3. *Cloud-Edge Collaboration: Balancing Elasticity and Real-Time*

The future scientific visualization system will adopt a "cloud-edge-end" collaborative architecture: edge nodes are responsible for real-time data preprocessing and local low-latency visualization, cloud computing platforms are responsible for large-scale data storage, complex analysis, and long-term storage tasks, and terminal devices (such as user mobile phones, tablets, and VR headsets) are seamlessly connected to cloud edge services through 5G/6G networks to achieve the unity of "real-time interaction + in-depth analysis." For example, in the construction of smart cities, environmental sensors (edge nodes) distributed throughout the city monitor air quality, noise level, and other data in real time and generate a simple pollution distribution heat map locally for citizens to view. After the key data is summarized in the cloud, a city-scale environmental quality trend report is generated through AI analysis and displayed to decision makers through a 3D visualization platform (such as a digital twin city) to support refined governance.

11.5. Interdisciplinary Integration and Domain Specificity: Deepening from "Universal Tools" to "Precision Solutions"

The future of scientific visualization is not only the upgrading of technology but also the deep integration with the needs of specific subjects — different scientific fields (such as life science, earth science, and material science) have unique data characteristics, scientific problems in science, and user community; a universal visual instrument cannot meet all the needs. As a result, domain-specific visualization systems will become mainstream, providing "accurately matched" visual solutions for specific disciplines through customized data adaptation, interaction design, and cognitive logic.

11.5.1. *Integration of Interdisciplinary Methodologies*

Scientific visualization will be deeply bound to the theoretical framework and research methods of specific subjects. For example:

- In quantum physics, visualization needs to combine the mathematical foundations of quantum mechanics (such as wave functions and arithmetic) and present non-classical properties of microscopic particles through metaphors, such as probability density graphs, energy-level transition animations, and other metaphors.
- The graph of a network of food webs and the graph of the thermal distribution of a species.
- In materials science, visualize the microstructure (e.g., crystal defects and interface structure) and macroscopic properties of materials (e.g., strength and conductivity) that need to be correlated with performance parameters through 3D atomic models.

11.5.2. *Development of Domain-Specific Tools*

Research institutions and enterprises will develop dedicated visualization software or platforms to address the pain points of a particular discipline. For example:

- "Single cell analysis visualization platform" in the field of life science, integrating the functions of single cell RNA sequencing data dimensionality reduction (such as t-SNE, UMAP), clustering (such as Leiden algorithm), and trajectory inference (such as Monocle algorithm), to visually display cell differentiation paths through three-dimensional scatter plots and branch trajectories;
- A "geo-hazard risk assessment system" in the field of geoscience, which integrates geological structural data (such as fault distribution), meteorological data (such as rainfall), and historical disaster records, and assists governments in disaster prevention planning through risk thermal maps and scenario simulations;
- A "digital twin platform for aero engines" in engineering that combines real-time sensor data (e.g., temperature and vibration) with physical models (e.g., CFD simulation and structural mechanics analysis) to monitor engine conditions and predict potential failures in real time through a dynamic visualization interface.

11.5.3. *Integration of Scientific Visualization and Science Education*

Scientific visualization is not only a tool of scientific research but also an important carrier of scientific education. In future educational scenarios, visualization technology will design immersive, gamified learning experiences based on the cognitive characteristics of students of different ages. For example:

- Primary and secondary science classes demonstrate abstract concepts such as protostructure and cell division through AR/VR technology, allowing students to "operate" molecular models or "walk into" the human digestion system.
- Professional courses in higher education (e.g., quantum mechanics and fluid mechanics) use interactive visualization software (e.g., Jupyter Notebook integrated visualization plugin) to help students understand complex mathematical formulas and physical processes.

- Volkswagen Science Fair aims to promote public interest in and participation in science through immersive exhibitions, such as “Digital Earth” and “Quantum Garden.”

The future of scientific visualization is “the universal language of cognitive science.”

The convergence of cutting-edge technologies is driving the transformation of scientific visualization from a “data presenter” to an “intelligent cognitive partner.” Future scientific visualization will have the following key characteristics:

- **More intelligent:** AI technology is integrated throughout the entire process, from data preprocessing to pattern discovery and visualization generation, automatically extracting key information, recommending optimal display solutions, and aiding scientific discoveries.
- **More immersive:** VR/AR/MR and multi-modal interaction technologies allow users to explore scientific phenomena “immersive,” deepening cognition through natural interaction and sensory synergy.
- **More powerful:** Cloud computing and edge computing provide flexible computing support for real-time rendering of large-scale data and collaborative analysis on a global scale.
- **More accurate:** Interdisciplinary integration and domain-specific design make visualizations deeply adapted to the specific problem in science, becoming the “exclusive cognitive instrument” for all fields.

The goal of scientific visualization is to make complex data and abstract scientific principles visible, perceptible, and understandable, so as to promote breakthroughs in scientific research, innovation of engineering technology, and improvement of public scientific literacy. In the process, it is not only a carrier of technology but also a bridge between human wisdom and natural mysteries. Just as Galileo opened the door to modern astronomy, scientific visualization may open a door to deeper scientific understanding.

Chapter 12

Training and Career Possibilities: "Breaking the Circle" and "Deepening" of Scientific Visualization

12.1. Introduction: When Technological Innovation Meets the Need for Human Resources — "Circle-breaking Power" and "Deep Plowing Path" in the Field of Scientific Visualization

In the new era of scientific visualization toward intelligence and integration, the rapid iterations of technologies, immersive scenarios built by virtual/augmented reality and real-time dynamic mapping enabled by digital twins are reshaping the boundaries and implications of this field with unprecedented force. Where scientific visualization might have been defined as a "scientific aid" or "data representation," it has grown to encompass, an "interdisciplinary hub" for engineering design, education communication, public science popularization, and even industrial innovation — from intelligent interpretation of genetic sequencing data to immersive decision support for climate change simulations. From the accurate diagnosis of medical imaging to the real-time visualization monitoring of smart city systems, scientific visualization talents are active in frontier laboratories for basic research, command centers for engineering projects, interactive screens in educational classrooms, and even science and technology exhibition halls for the public.

This shift in role stems from the evolution of technical tools (visual systems have been upgraded from static chart generators to cognitive intelligent companions). It is also due to the complexity and diversity of the problem in science. As a result, there is an urgent need for scientific visualization talents. They need to not only master traditional data processing, graphics programming, and visual design skills but also understand the research logic and data characteristics of different disciplines (such as life science, earth science, and physical science). Not only do you need to be familiar with interactive techniques and front-end development tools, but you also need to have the "translational ability" to translate complex scientific concepts into intuitive visual language. More importantly, they need to find a balance between "technical feasibility" and "user needs" — where they can leverage cutting-edge technologies such as AI-driven design generation to meet user demands effectively. VR immersion enhances the scientific and innovative nature of visualization and can design truly valuable visualization solutions with a "user-centered" mindset (such as the precise analysis needs of researchers and the easier-to-understand needs of the public).

However, there are still many challenges and opportunities in the field of scientific visualization. Design focuses on aesthetics and natural science focuses on theories, leading to a "knowledge fault line" in the development of composite talents — "cross-boundaries" who understand science, technology, and design remain a scarce resource. On the other hand, the rapid iterations of technology (such as breakthroughs in big model technology and the popularity of new interactive devices) require continuous learning and adaptation by practitioners. The diversity of career paths — from visualization researchers at scientific institutions to product designers at companies, from popular science developers in education to policy support specialists in the public sector — requires clearer guidance for growth.

It is these challenges and opportunities that push the cultivation and career of scientific visualization into a new stage of "breaking the circle" and "deepening the cultivation." "Breaking the circle" means to break through the traditional disciplinary boundaries and professional boundaries. Through interdisciplinary education, industrial and research cooperative training models, we will build a composite talent pipeline that combines scientific literacy, technical ability, and design thinking. "Deep plowing" requires practitioners to work in specific areas such as scientific visualization. By accumulating in-depth expertise in earth science visualization and engineering simulation visualization, one can become a "vertical expert" capable of solving complex problems. At the same time, exploring the deep integration of emerging technologies such as generative AI and neural rendering with scientific visualization can open up new possibilities for career development.

This chapter focuses on the key proposition of talent training and career development in the field of scientific visualization and systematically analyzes the core competencies, training paths, and multiple career possibilities for future talents:

- **Core competencies spectrum:** From "hard skills" (e.g., programming language Python/JavaScript, data visualization library D3.js/Three.js, and scientific data processing tools Pandas/NetCDF) to "soft skills" (such as interdisciplinary communication skills) to "emerging technology sensitivity" (such as artificial intelligence aided design, virtual reality interactive logic, and digital twin system construction), what knowledge and ability scientific visualization talents need to have to deal with future challenges.
- **Innovation in training models:** This explores how universities and research institutions can design interdisciplinary curricula (such as "Science + Computer + Design" joint training programs), collaboratives practices between industry and academia (such as jointly developing visualization projects with hospitals, enterprises, and museums) and online education platforms (such as MOOCs and workshops) to develop composite talents adapted to the needs of the times, and examines how businesses and social organizations support the continued growth of practitioners through mentoring, project-based learning, and skills certification systems.
- **Diversity of career paths:** This is from Visual Investigator in scientific institutions (focusing on visual explorations of cutting-edge problem in science), technical experts in engineering enterprises (providing visualization support for product design and industrial simulation), and science developers in education (creating scientific visualization content for the public), to the public sector policy support experts (promoting the application of scientific visualization in smart cities, environmental protection and other fields), to analyze the core responsibilities, skills requirements, and development prospects of different career directions, to provide clear career planning reference for practitioners.

- **Practice cases that break boundaries and cultivate:** This uses real-life examples (such as how an interdisciplinary team produced a "science-design-programming" composite talent to lead major scientific visualization projects, or a practitioner has become a specialist in medical visualization and opened up a new market for digital therapy), showing how scientific visualization professionals can expand their boundaries in the "breaking of the circle," establish barriers in the "deep cultivation," and ultimately achieve a win-win situation for both individuals and the field.

The future of scientific visualization is essentially defined by the creativity and professionalism of people. When a wave of technology pushes the field to continually "break circles" and expand its boundaries, when a dedication to cultivation allows practitioners to build irreplaceable advantages in a particular field, scientific visualization talents will become the "key link" connecting scientific discovery, technological innovation, and human cognition. They are not only the creators of visualization tools but also the disseminators of scientific knowledge, the solvers of complex problems, and the companions of future scientific exploration. The discussion in this chapter provides a guide for any reader (whether student, practitioner, or educator) who is interested in the field of scientific visualization. We help more people find their own path of "breaking circles" and "cultivating" in this dynamic field (Zhang *et al.*, 2021; Lee *et al.*, 2022).

12.2. Core Competencies of Future Talent: Upgrading from "Single Skills" to "Composite Literacy"

The core competence of scientific visualization professionals lies in their ability to bridge the "problem in science," "data technologies," and "customer needs." The biological background of biological data also requires mastering visualization techniques (such as 3D modeling, interaction design, and algorithm development) and the ability to transform complex information into intuitive expression (such as user experience design and scientific narrative logic). Specifically, future talents need to have the following four core competence dimensions (Liu *et al.*, 2023; Zhao *et al.*, 2023).

12.2.1. *Scientific Literacy: Understanding the "Why" Behind the Data*

Scientific visualization is not "visualization for visualization's sake" but serves specific scientific goals (such as verifying theoretical hypotheses, discovering hidden rules, and assisting decision-making). Therefore, talents need to have solid scientific knowledge of the field and be able to understand the source, meaning, and scientific value of data. For example:

- In life sciences, you must know the biological significance of gene expression data (e.g., which genes are associated with disease) and the dynamic properties of protein structure (e.g., the effect of conformational changes on function).
- In the field of geosciences, it is necessary to grasp the observational logic of meteorological data (e.g., the principles of satellite remote sensing) and the laws of movement of geological formations (e.g., the mechanisms of activity of plate boundaries).
- In engineering, you need to be familiar with the working principles of mechanical equipment (such as the heat cycle of aero engines) and the electrical properties of electronic chips (such as signal transmission delays and interference).

This scientific literacy not only helps people to "read the data correctly" but also leads them to focus on the key problem in science ("Which variables are most important to the goal?" and "How to visualize the cause-and-effect relationship between variables?") (Hansen, *et al.*, 2014; NASA, 2023).

12.2.2. *Technical Capabilities: Harnessing a Full-Link Tool Chain For "Data-Graphics-Interaction"*

The implementation of scientific visualization relies on the cooperation of multiple technological stacks. The person should master the whole process from data acquisition and processing to visual presentation and interactive development, which includes the following:

- **Data processing and analysis:** Familiar with Python (Pandas and NumPy), R, and other programming languages, mastering data cleansing (e.g., missing value processing and anomaly detection), statistical analysis (e.g., correlation analysis and cluster analysis), and feature engineering (e.g., deflection and feature selection) techniques; understand the common formats for scientific data (e.g., NetCDF, HDF5, FITS) and storage methods (e.g., distributed database and cloud storage).

- **Visualization tools and frameworks:** Proficient in professional visualization tools such as ParaView, VTK for scientific computational data, Unity3D/Unreal Engine for immersive visualization, general-purpose visualization libraries (e.g., D3.js, Matplotlib, and Plotly for web-side interactive graphs), and 3D modeling software (e.g., Blender and Maya for complex scene building); master the fundamentals of graphics (e.g., OpenGL/DirectX rendering pipelines and shader programming) to achieve customized visual effects.
- **Interaction and front-end development:** Familiar with front-end technologies such as HTML/CSS/JavaScript, master interaction design principles (e.g., user path optimization and feedback mechanism design), and able to develop responsive visual interfaces (e.g., web applications that support multi-device access); understand the foundations of artificial intelligence (e.g., machine learning model calls and natural language processing interfaces) to integrate intelligent analytical functions (e.g., automatically recommending visualization schemes) (Kumar *et al.*, 2022; Wang *et al.*, 2023).

12.2.3. *Design Thinking: Balancing "Scientific Rigor" with "User-Friendly"*

Scientific visualization requires both "scientific accuracy" and "user understandability" — too much technical flair can lead to information overload (e.g., too many distracting colors and animations), while over-simplification can lead to the loss of critical scientific details. Therefore, talents need to have the ability to think designedly, which can be expressed as follows:

- **Visual metaphorical design:** This chooses appropriate visual coding (e.g., color to map temperature and flow lines to represent fluid flow) and avoids misleading design (e.g., incorrect color gradient direction);

lowers the cognitive threshold for users through graphical metaphors, such as using "garden" to mean an ecosystem and "city" to mean a data network.

- **Interaction logic optimization:** This designs intuitive interactive actions (such as mouse hovering to display details and sliding to adjust parameters) to ensure that users can access information autonomously through the "exploration-discovery" process; multiple view interactions (such as displaying time series graphs and spatial distribution graphs simultaneously) help users establish data associations.
- **User needs insights:** This targets different user groups (e.g., researchers, engineers, and the public) and adjusts their visualization strategy — research users may need high-precision, multivariate complex graphs, while the public focuses more on fun, storytelling, and simplicity (e.g., using animation to tell the process of scientific discovery).

12.2.4. *Interdisciplinary Collaboration and Communication Skills: Becoming a "Translator" and a "Coordinator"*

Scientific visualization projects usually involve multidisciplinary teams (e.g., scientists, engineers, designers, and product managers), with people acting as "bridges" across disciplines:

- Communicating with scientists with an accurate understanding of their research objectives (e.g., "to validate predictions of a physical model"), data characteristics (e.g., the data are high-dimensional time series and require sub-dimensional processing), and scientific constraints (e.g., "the physical unit of the original data must be retained");
- Ability to translate scientific needs into technologically achievable functions when working with engineers (e.g., "real-time rendering of large-scale particle data is required, GPU acceleration is recommended");
- Be able to balance scientific rigor with visual beauty when working with designers (e.g., "color schemes need to conform to scientific practices while avoiding visual fatigue");

- The ability to translate complex scientific concepts into plain language when presented to the public (e.g., "the Earth breathes" to describe climate change and "digital twins" to explain industrial simulations).

12.3. Talent Development System: From "Interdisciplinary Curriculum" to "Practical Training"

At present, universities and research institutions around the world have begun to explore the scientific visualization-related personnel training model. The core is to cultivate interdisciplinary talents with scientific depth and technical breadth through interdisciplinary courses and practice-oriented training programs. Typical training paths can be divided into "academic" (for research and higher education) and "applied" (for engineering and industry), but both emphasize the following key elements.

12.3.1. *Cross-Disciplinary Curricula: Breaking Down Barriers Between "Science" and "Technology"*

Scientific visualization courses need to integrate the core content of natural science (such as physics, biology, and earth science), engineering science (such as computer science and electronic information engineering), and humanities (such as design and communication). Typical course modules include the following:

- **Basic Scientific Modules:** Selected according to the sub-modules (e.g., cell biology and molecular biology), earth scientific visualization requires a course in meteorology and geophysics to help students understand the scientific implications of field data.
- **Technical core modules:** These cover programming languages (Python and C++), data structure and algorithm, computer graphics (three-dimensional modeling and rendering technology), visualization principles (color theory and visual perception), human–computer interaction design (user experience and interaction logic), and lay the technical implementation ability.

- **Tools and methodology module:** This teaches the use of professional visualization tools (e.g., ParaView, VTK, and Unity3D), as well as scientific data processing methods (e.g., statistical analysis and machine learning basics) and visual design methods (e.g., infographic design and narrative visualization).
- **Cross-disciplinary seminar modules:** Through case studies (e.g., "How to visualize a quantum computation process" and "How to design a public picture of the spread of influenza"), team projects (such as working with a scientific team to visualize a real scientific data task) develop students' ability to solve complex problems (Lee *et al.*, 2022; Zhang *et al.*, 2021).

12.3.2. *Practice-Oriented Training Projects: From "The Classroom" to "Real Scenarios"*

Theoretical knowledge needs to be translated into competence through practice, so talent development needs to emphasize a "project-driven" learning model:

- **Course experiments:** These include mastering basic tools and technical processes through small projects such as "t-SNE mapping of single-cell RNA sequencing data in Python" and "developing dynamic maps of COVID-19 data based on D3.js."
- **Research collaborations:** These include participating in practical projects in universities and research institutions (e.g., visualizing galactic evolution data for astronomical teams and creating dynamic climate maps for environmental science teams) to practice data adaptation, visualization, and interdisciplinary communication in a real problem in science.
- **Competitions and open-source contributions:** Students are encouraged to participate in international visualization competitions (e.g., IEEE VIS and SciVis Contest) or contribute code to open-source visualization tools (e.g., D3.js and VTK) to learn cutting-edge technologies and best practices in the community.
- **Corporate Internship:** Internships in a technology company (e.g., in the scientific data department of a visual software developer or an Internet company), an engineering company (e.g., in the digital

department of an intelligent manufacturing company), or a science institute (e.g., in a science museum or a natural history museum) to understand the needs and career scenarios of the industry.

12.3.3. *Continuous Learning and Lifelong Education: Rapid Iterations of Adaptation Technologies*

Scientific visualization technologies, such as artificial intelligence-generated visualization and immersive experience in the meta-universe, are developing rapidly. Students and practitioners must take online courses such as Coursera's "Data Visualization Course," edX, industry conferences (e.g., IEEE VIS and China Visualization & Visualization), and tech communities (e.g., GitHub and Stack Overflow) to keep track of the latest developments and the body of knowledge.

12.4. Career Development Paths: Multiple Possibilities for "Breaking Circles" and "Deep Cultivation"

The career direction of scientific visualization talents is highly flexible, not only in the traditional scientific research and engineering fields but also in the emerging science, education, and industrial tracks. Here are a few typical career paths.

12.4.1. *Scientific Research and Academic Direction: To Become a "Scientific Visualization Specialist"*

- **Universities and research institutes:** These include teaching in computer science, data science, science communication, and other related departments, offering scientific visualization courses to cultivate the next generation of talent; or join interdisciplinary labs (e.g., computational biology and geo-information science), focus on scientific visualization methodologies (e.g., multi-scale data fusion algorithms and immersive scientific experience design), and publish high-level papers (e.g., IEEE Transactions on Visualization and Computer Graphics).
- **Research institutions:** In national laboratories (e.g., the Chinese Academy of Sciences and the United States Department of Energy

National Laboratory), professional research institutes (such as the Max Planck Institute in Germany) work as visualization engineers or researchers, providing customized visualization support for major scientific research projects (such as particle physics experiments and climate change simulations) and promoting the visualization of scientific discoveries.

12.4.2. *Engineering and Industry Direction: Becoming a "Digital Solutions Expert"*

- **Technology companies:** In visual software companies (such as Kitware and Tableau) and interconnected companies (such as Google's Earth Engine team or Alibaba Cloud's digital twin division) engages in product research and development, develops general-purpose or industry-specific visualization tools (such as industrial equipment monitoring platforms and financial data visualization large screens), and solves enterprise-level data presentation and analysis needs.
- **Engineering:** In smart manufacturing (e.g., aero engine design firms) and energy (e.g., geological modeling teams at oil exploration firms), he works as a visual engineer in industries such as architectural design institute's BIM visualization department. He assists with engineering decisions such as optimizing assembly line layout and predicting building energy consumption through 3D modeling, dynamic simulation, and interactive design.
- **Consulting and technical services:** As an independent consultant or joining a professional consulting firm (such as Accenture's digital transformation team), they provide scientific visualization consulting services to enterprises (such as "how to improve plant operation efficiency through visualization" and "how to design energy project reporting charts for investors") to solve data communication problems in specific scenarios.

12.4.3. *Popularization of Science and Education: Becoming a "Bridge for Science Communication"*

- **Science and Technology Museum:** It acts as scientific visualization designer or science curator, develops immersive science exhibitions (such as "Digital Earth" and "Quantum World"), transfers scientific knowledge to the public through interactive installation, moving image, and game design, and improves the scientific literacy of the people.

- **Educational institutions:** In schools (e.g., science teachers), universities (e.g., general studies course developers), or online (e.g., Coursera's scientific visualization course developers), integrating visualization techniques into the teaching process (e.g., using animations to illustrate cell division and interactive graphs to illustrate chemical reactions) helps students understand scientific concepts more intuitively.
- **New Media and Content Creation:** As a science blogger, popular science video producer, or popular science writer, utilize visualization tools such as After Effects animation and Blender 3D modeling to create interesting scientific content (e.g., "Take you around human organs with VR" and "interpret Nobel Prize discoveries with data graphs") and expand the impact of science communication through social media (e.g., TikTok, B station, and YouTube).

12.4.4. *Entrepreneurial and Industrial Innovation Direction: To Become a "Scientific Visualization Trail Blazer"*

- **Startups:** These include founding a visualization technology company based on market needs (e.g., the digital visualization needs of SMEs and customized tools gaps for specific industries), developing visualization solutions for vertical fields (e.g., crop growth monitoring platform in agriculture, surgical planning visualization system in medical field), and promoting the commercialization of scientific visualization technology.
- **Open-source community contributors:** These include developing open-source visualization tools (such as improving D3.js's scientific charting module, VTK), to build an open tool ecosystem, to reduce the barriers to the use of scientific visualization, and to obtain technical influence and business opportunities through community cooperation.
- **The future mission of scientific visualization talents:** "Let science be seen, let knowledge be understood."

The essence of scientific visualization talents is the "translator" of science and technology, the "decipherer" of complex information, and the "extender" of cognitive boundaries. They need to not only master the techniques of "how to do visualization" but also think about "why visualization is needed," "for whom visualization is necessary," and "how to promote scientific progress and social development through visualization." With the deepening of trends such as artificial intelligence, virtual reality, and interdisciplinary research, the demand for talent in this field will continue to grow. The boundaries of their professions will also expand — from traditional "tool developers" to "problem solvers," from "researcher supporters" to "science communicators," and from "technology implementers" to innovation leaders.

For learners and practitioners interested in pursuing a career in this field, there are opportunities and challenges ahead: deepening foundational skills in technology and science while fostering interdisciplinary perspectives and innovative thinking. We must not only pay attention to the cutting-edge dynamics of technology but also adhere to the original intention of "user-centered" design. As visualization guru Edward Tufte said, "Good visualization is not the decoration of data, but the carrier of ideas." Scientific visualization talents are the shapers of this career. They will make the mysteries of science accessible and enlighten more people's understanding through visual language.

Bibliography

Amunts, K., Lippert, T., & Zilles, K. (2021). Towards a comprehensive atlas of the human brain. *Nature Reviews Neuroscience*, 22(3), 187–200. https://doi.org/10.1038/s41583-020-00424-6.

Andrienko, G., Andrienko, N., Miksch, S., & Thonnard, S. (2020). *Interactive Visual Analytics for Spatial and Temporal Data*. Springer, London. https://doi.org/10.1007/978-3-030-47442-8.

ANSYS. (2023). CFD and FEA simulation software for engineering. https://www.ansys.com.

Berman, F., Fox, G., & Hey, T. (2022). *Grid Computing: Making the Global Infrastructure a Reality*. Wiley-Blackwell, New Jersey.

Bertin, J. (1967). *Sémiologie Graphique: Les Diagrammes, Les Réseaux, Les Cartes* [Graphic semiology: Diagrams, networks, maps]. Gauthier-Villars. (English translation: 1983, Graphics and Graphic Information Processing).

Borgman, C. L. (2019). The lives and after lives of data. *Harvard Data Science Review*, 1(1). https://doi.org/10.1162/99608f92.9a36bdb6.

Brandes, U. & Raab, J. (2021). Network visualization: A comprehensive survey. *IEEE Transactions on Visualization and Computer Graphics*, 27(5), 1–25.

Burch, M. & Weiskopf, D. (2022). Interactive generative visualization for science communication. *Computer Graphics Forum*, 41(3), 1–18.

Card, S. K., Mackinlay, J. D., & Shneiderman, B. (2022). *Readings in Information Visualization: Using Vision to Think* (2nd ed.). Morgan Kaufmann, San Francisco.

CDC. (2023). COVID-19 Data Tracker. Centers for Disease Control and Prevention. https://covid.cdc.gov/covid-data-tracker.

CERN. (2021). Large Hadron Collider (LHC) data and experiments. https://home.cern/science/experiments.

Cesium. (2022). CesiumJS – Open Source JavaScript Library for World-Class 3D Globes. https://cesium.com/platform/cesiumjs/.

Clark, J. S., *et al.* (2022). Long-term ecological monitoring with remote sensing and ground data. *Ecological Applications*, 32(3), e2545. https://doi.org/10.1002/eap.2545.

Cleveland, W. S. (2021). *The Elements of Graphing Data* (Revised ed.). Hobart Press, New Jersey.

Das, S., *et al.* (2023). Longitudinal imaging biomarkers in Alzheimer's disease: From amyloid to atrophy. *The Lancet Neurology*, 22(4), 312–325.

Dykes, J., MacEachren, A. M., & Kraak, M. J. (2020). *Exploring Geovisualization* (2nd ed.). Pergamon Press, Oxford.

ENCODE Project Consortium. (2022). Expanded encyclopedias of DNA elements in the human genome. *Nature*, 603(7903), 1–10. https://doi.org/10.1038/s41586-022-04495-4.

Esteva, A., Kuprel, B., Novoa, R. A., Ko, J., Swetter, S. M., Blau, H. M., & Thrun, S. (2021). Dermatologist-level classification of skin cancer with deep neural networks. *Nature*, 601(7893), 101–108. https://doi.org/10.1038/nature21056.

European Space Agency (ESA). (2022). Earth Observation Multimedia Centre: The Phi Experience. ESA-ESRIN.

Eyring, V., Bock, L., Lauer, A., Righi, M., Schlund, M., Andela, B., ... & Zechlau, S. (2021). Earth System Model Evaluation Tool (ESMValTool) v2.0 – technical overview. *Geoscientific Model Development*, 14(3), 1319–1352. https://doi.org/10.5194/gmd-14-1319-2021.

FlyEM Project. (2023). Complete Connectome of the Adult Fruit Fly Brain. HHMI Janelia Research Campus. https://www.janelia.org/project-team/flyem.

Fortunato, S. (2020). Community detection in graphs. *Physics Reports*, 894, 1–80.

Getoor, L. & Taskar, B. (Eds.). (2007). *Introduction to Statistical Relational Learning*. MIT Press. https://doi.org/10.7551/mitpress/7432.001.0001.

Gershon, N. & Page, W. (2020). *What Is Data? A Multidisciplinary Perspective*. MIT Press, Cambridge, MA.

Gogtay, N., *et al.* (2020). Dynamic mapping of human cortical development during childhood through early adulthood. *PNAS*, 117(40), 24911–24920. https://doi.org/10.1073/pnas.1912204117.

Goodfellow, I., Bengio, Y., & Courville, A. (2022). *Deep Learning* (2nd ed.). MIT Press, Cambridge, MA.

Goodfellow, I., *et al.* (2020). Generative adversarial networks. *Communications of the ACM*, 63(11), 1–15.

Gupta, R., Singh, S. K., & Rahman, A. (2021). Application of geospatial technology in sustainable mineral resource management. *Resources Policy*, 74, 102392. https://doi.org/10.1016/j.resourpol.2021.102392.

Hamilton, J. P. (2022). Functional connectivity in depression: Insights from dynamic fMRI. *Biological Psychiatry*, 91(5), 432–440.

Han, J., Pei, J., & Kamber, M. (2020). *Data Mining: Concepts and Techniques* (4th ed.). Morgan Kaufmann, San Francisco.

Hansen, A., MacKay, N. J., & Ranganathan, J. (2023). Communicating climate change risks through visualization. *Environmental Communication*, 17(1), 45–60. https://doi.org/10.1080/17524032.2022.2145637.

Hansen, C. D., Chen, M., Johnson, C. R., Kaufman, A. E. & Hagen, H. (2014). *Scientific Visualization: Uncertainty, Multifield, Biomedical, and Scalable Visualization*. Springer, London.

Heer, J. & Shneiderman, B. (2021). Interactive dynamics for visual analysis. *Queue*, 19(2), 1–30. https://doi.org/10.1145/3447927.

Heer, J., Bostock, M., & Ogievetsky, V. (2021). A tour through the visualization zoo. *Communications of the ACM*, 64(12), 50–61. https://doi.org/10.1145/3458778.

Hegarty, M. (2020). The cognitive science of visualization. *Current Directions in Psychological Science*, 29(4), 350–355. https://doi.org/10.1177/0963721420915839.

Helge, K. (2021). *Quantum Generations: A History of Physics in the Twentieth Century*. Princeton University Press, New Jersey.

Hertzmann, A. (2021). Creative AI and procedural generation in visualization. *IEEE Computer Graphics and Applications*, 41(4), 22–33.

Hochreiter, S. & Schmidhuber, J. (2021). Long short-term memory. *Neural Computation*, 9(8), 1735–1780. https://doi.org/10.1162/neco.1997.9.8.1735.

Iliinsky, N., Steele, J. (2011). *Designing Data Visualizations: Representing Informational Relationships*. O'Reilly Media. Sebastopol, CA, United States.

ILISI. (2021). Ethical Guidelines for Medical Data Visualization. International Life Sciences and Informatics Institute. https://www.ilisi.org/ethics.

Inselberg, A. (2009). *Parallel Coordinates: Visual Multidimensional Geometry and Its Applications*. Springer, London.

IPCC. (2021). Sixth Assessment Report (AR6) – Climate Change 2021: The Physical Science Basis. Intergovernmental Panel on Climate Change. https://www.ipcc.ch/report/ar6/wg1/.

Jack, C. R., *et al.* (2020). Tracking Alzheimer's disease progression with multimodal imaging. *Brain*, 143(6), 1742–1758.

Jacomy, M., Venturini, T., Heymann, S., & Bastian, M. (2021). ForceAtlas2: A continuous graph layout algorithm for handy network visualization. *Journal of Graph Algorithms and Applications*, 25(1), 1–25.

Javed, W. & Elmqvist, N. (2016). Exploring the design space of composite visualization. *IEEE Transactions on Visualization and Computer Graphics*, 22(1), 309–318. https://doi.org/10.1109/TVCG.2015.2467736.

Jiang, W., Zhou, Z., Li, G., Yin, W., *et al.*, (2023). Mapping the evolution of regional brain network efficiency and its association with cognitive abilities during the first twenty-eight months of life. *Developmental Cognitive Neuroscience*, 63, 101284. https://doi.org/10.1016/j.dcn.2023.101284.

Johnson, A. & Roussou, M. (2022). Immersive analytics and VR for scientific exploration. *IEEE Computer Graphics and Applications*, 42(1), 56–65. https://doi.org/10.1109/MCG.2021.3134567.

Johnson, G. T., *et al.* (2020). Molecular visualization in structural biology. *Nature Methods*, 17(5), 431–440.

Jordan, M. I. & Mitchell, T. M. (2021). Machine learning: Trends, perspectives, and prospects. *Science*, 372(6540), 339–342. https://doi.org/10.1126/science.aaz9772.

Keim, D. A., *et al.* (2021). Visual analytics for high-dimensional data. In *IEEE VAST* (pp. 45–54).

Keim, D. A., Mansmann, F., Schneidewind, J., Thomas, J., & Ziegler, H. (2021). Visual analytics: Scope and challenges. In *IEEE Symposium on Visual Analytics Science and Technology (VAST)* (pp. 1–18). https://doi.org/10.1109/VAST50702.2021.9605598.

Kim, J., *et al.* (2022). Multimodal brain imaging fusion in epilepsy and psychiatric disorders. *NeuroImage: Clinical*, 35, 103087.

Kitchin, R. (2022). *Data Science and the Formation of Knowledge*. MIT Press, Cambridge, MA.

Klein, T., *et al.* (2022). Interactive visualization design principles for neuroscience education. *Frontiers in Neuroinformatics*, 16, 856791.

Kosara, R. (2021). Visualization criticism – The missing link between theory and practice. *IEEE Computer Graphics and Applications*, 41(4), 72–80. https://doi.org/10.1109/MCG.2021.3072415.

Kuhn, T. S. (2022). *The Structure of Scientific Revolutions* (2nd ed.). University of Chicago, Chicago Press, Chigago (Original work published 1962).

Kumar, P., Chatterjee, C., & Sharma, A. (2022). Visualization techniques for climate and disaster risk analysis. *Natural Hazards Review*, 23(1), 04021054. https://doi.org/10.1061/(ASCE)NH.1527-6996.0000512.

Lam, H., Bertini, E., Isenberg, P., Plaisant, C., & Carpendale, S. (2020). Empirical studies in information visualization: Seven scenarios. *IEEE Transactions on Visualization and Computer Graphics (TVCG)*, 26(8), 2622–2637. https://doi.org/10.1109/TVCG.2019.2934522.

Laumann, T. O., *et al.* (2021). Mapping the functional connectome in healthy adults. *Cerebral Cortex*, 31(3), 1771–1790.

LeCun, Y., Bengio, Y., & Hinton, G. (2022). Deep learning for perception and understanding. *Nature*, 604(7906), 440–450. https://doi.org/10.1038/s41586-022-04580-y.

Lee, J. & Kang, H. (2021). Advanced visualization of histopathological data in oncology. *Journal of Pathology Informatics*, 12, 12.

Lee, J., Kim, S., & Park, E. (2022). Design thinking in scientific visualization education. *International Journal of Technology and Design Education*, 32(2), 889–905. https://doi.org/10.1007/s10798-021-09710-3.

Li, X., Peters, T., & Zhang, Y. (2022). 3D medical imaging in surgical planning: A review. *Medical Image Analysis*, 75, 102256.

Liu, Y., *et al.* (2023). Virtual and augmented reality in neurosurgical training and brain exploration. *Nature Digital Medicine*, 6(1), 28.

Liu, Y., Kumar, P., Zhang, X., & Shi, P. (2020). Big Earth Data supports sustainable development goals: A review. *International Journal of Applied Earth Observation and Geoinformation*, 89, 102068. https://doi.org/10.1016/j.jag.2020.102068.

Liu, Y., Kumar, P., Zhang, X., & Shi, P. (2023). Big Earth Data and AI-driven visualization for sustainable development. *International Journal of Applied Earth Observation and Geoinformation*, 115, 103089. https://doi.org/10.1016/j.jag.2023.103089.

Liu, Z. & Chen, M. (2021). Scientific visualization: Principles and applications. *Journal of Visualization*, 24(3), 455–475. https://doi.org/10.1007/s12650-021-00783-z.

Liu, Z., Nersessian, N., & Stasko, J. (2008). Distributed Cognition as a Theoretical Framework for Information Visualization. in *IEEE Transactions on Visualization and Computer Graphics*, 14(6), 1173–1180. doi: 10.1109/TVCG.2008.121.

LSST Science Collaboration. (2023). Legacy Survey of Space and Time (LSST) data products. https://www.lsst.org/scientists/dataproducts.

Lynch, C. (2020). *Big Data and the Future of Scientific Discovery*. Springer, London.

Ma, Z. & Liu, J. (2021). Challenges in multimodal scientific data fusion. *IEEE Transactions on Visualization and Computer Graphics*, 27(2), 1234–1246. https://doi.org/10.1109/TVCG.2021.3054567.

Maaten, L. van der & Hinton, G. (2008). Visualizing data using t-SNE. *Journal of Machine Learning Research*, 9, 2579–2605.

Maaten, L. van der & Hinton, G. (2021). Visualizing data using t-SNE. *Journal of Machine Learning Research*, 22(1), 1–40.

MacEachren, A. M. (2021). *How Maps Work: Representation, Visualization, and Design* (2nd ed.). Guilford Press, New York.

Mackinlay, J. (1986). Automating the design of graphical presentations of relational information. *ACM Transactions on Graphics (TOG)*, 5(2), 110–141.

Mackinlay, J., Hanrahan, P., & Stolte, C. (2020). Show me: Automatic presentation for visual analysis. *IEEE Transactions on Visualization and Computer Graphics*, 26(1), 1–15. https://doi.org/10.1109/TVCG.2020.2971234.

Mahmood, A. R., Punni, S., & Aref, W. G. Spatio-temporal access methods: a survey (2010–2017). (2019). *Geoinformatica*, 23, 1–36. https://doi.org/10.1007/s10707-018-0329-2.

Munzner, T. (2022). *Visualization Analysis and Design* (2nd ed.). CRC Press, Boca Raton, FL.

NASA Earthdata. (2022). Multimodal Earth observation data for climate and health. https://earthdata.nasa.gov.

NASA. (2022). AI and visualization in Earth and space science. https://www.nasa.gov/.

NASA. (2022). Earth Observing System Data and Information System (EOSDIS). https://earthdata.nasa.gov/.

NASA. (2023). Global Climate Change: Vital Signs of the Planet. https://climate.nasa.gov/.

NASA. (2023). Scientific visualization in Earth and space science. https://www.nasa.gov/.

Nielsen, J. (1994). *Usability Engineering*. Morgan Kaufmann, San Francisco.

OGC. (2021). CF-Conventions for climate and forecast metadata. Open Geospatial Consortium. https://cfconventions.org.

Paus, T., *et al.* (2021). Brain development and plasticity across the lifespan. *Annual Review of Neuroscience*, 44, 215–238.

Peters, T. M., *et al.* (2022). Visualization in clinical oncology: From imaging to intervention. *Radiology: Artificial Intelligence*, 4(2), e210210.

Pirolli, P. & Card, S. (2020). *Information Foraging Theory: Adaptive Interaction with Information*. Oxford University Press, Oxford.

Roussou, M. (2022). Immersive VR for scientific education and exploration. In *Handbook of Immersive Analytics* (pp. 123–145). Springer, London.

Sacha, D., Stoffel, A., Stoffel, F., Kwon, B. C., Ellis, G., & Keim, D. A. (2020). Role of interactive visualization in exploratory data analysis. *IEEE Transactions on Visualization and Computer Graphics*, 26(1), 1–20. https://doi.org/10.1109/TVCG.2019.2934689.

Sedlmair, M., Isenberg, P., & Burch, M. (2021). Towards a theoretical foundation of visualization for data science. *Computer Graphics Forum*, 40(3), 331–354. https://doi.org/10.1111/cgf.14292.

Segel, E. & Heer, J. (2021). *Narrative Visualization: Telling Stories with Data*. MIT Press, Cambridge, MA.

Shapin, S. (2021). *The Scientific Life: A Moral History of a Late Modern Vocation*. University of Chicago Press, Chicago.

Sheppard, S. R. J. (2021). *Visualizing Climate Change: A Guide to Visual Communication of Climate Change and Developing Local Solutions* (2nd ed.). Routledge, Milton Park.

Shneiderman, B. (1997). Direct manipulation for comprehensible, predictable and controllable user interfaces. In *Proceedings of the 2nd International*

Conference on Intelligent User Interfaces (pp. 33–39). ACM. https://doi.org/10.1145/238218.238247.

Silver, D., Huang, A., Maddison, C. J., Guez, A., Sifre, L., Van Den Driessche, G., ... & Hassabis, D. (2021). Mastering the game of Go with deep neural networks and tree search. *Nature*, 529(7587), 484–489. https://doi.org/10.1038/nature16961.

Skarbez, R., Brooks, F. P., & Whitton, M. C. (2020). A survey of presence and related concepts. *ACM Computing Surveys*, 53(6), 1–39. https://doi.org/10.1145/3380909.

Slater, M. & Sanchez-Vives, M. V. (2016). Enhancing our lives with immersive virtual reality. *Frontiers in Robotics and AI*, 3, 74. https://doi.org/10.3389/frobt.2016.00074.

Smith, A. *et al.* (2021). Big data in life sciences: Challenges and visualization strategies. *Nature Computational Science*, 1, 412–420.

Smith, J. *et al.* (2020). High-throughput chemical screening: Data challenges and solutions. *Journal of Chemical Information and Modeling*, 60(5), 2345–2356. https://doi.org/10.1021/acs.jcim.9b00999.

Spence, I. (2020). *Information Visualization: Design for Interaction* (3rd ed.). Pearson, London.

Spence, R. (2021). *Information Visualization: Design for Interaction* (3rd ed.). Pearson, London.

Stanney, K. M., Fuchs, S., & Bowers, C. (2021). *Handbook of Virtual Environments: Design, Implementation, and Applications* (3rd ed.). CRC Press, Boca Raton, FL.

Sutton, R. S., & Barto, A. G. (2022). *Reinforcement Learning: An Introduction* (2nd ed.). MIT Press, Cambridge, MA.

Sweller, J. (2020). Cognitive load theory and educational technology. *Educational Technology Research and Development*, 68(1), 1–16. https://doi.org/10.1007/s11423-019-09704-3.

Tao, H., *et al.* (2021). Precision localization of epileptogenic zones using fused fMRI-EEG-DTI. *Epilepsia*, 62(8), 1889–1901.

Taylor, C. A., *et al.* (2021). Cardiovascular flow simulation for surgical planning. *Journal of Biomechanical Engineering*, 143(6), 061001.

Tolle, K. M., Tansley, D. S., and Hey, A. J. G. (2011). The Fourth Paradigm: Data-Intensive Scientific Discovery [Point of View]. in *Proceedings of the IEEE*, 99(8), 1334–1337. doi: 10.1109/JPROC.2011.2155130.

Tuckerman, M. E., *et al.* (2022). Quantum chemistry simulations at scale: Methods and applications. *Journal of Chemical Physics*, 156(4), 044101. https://doi.org/10.1063/5.0080123.

Tufte, E. R. (1983). *The Visual Display of Quantitative Information*. Graphics Press, Connecticut.

Tufte, E. R. (2021). *The Visual Display of Quantitative Information* (2nd ed.). Graphics Press, Connecticut.

Tufte, E. R. (2022). *Beautiful Evidence* (Anniversary ed.). Graphics Press, Connecticut.

Tversky, B. & Morrison, J. (2021). Visualizing science: Tools for scientific storytelling. *Cognitive Research: Principles and Implications*, 6(1), 1–15. https://doi.org/10.1186/s41235-021-00278-3.

Tversky, B., Morrison, J. B., & Betrancourt, M. (2020). Animation: Can it facilitate? *International Journal of Human-Computer Studies*, 144, 102499. https://doi.org/10.1016/j.ijhcs.2020.102499.

UNESCO. (2022). Recommendation on the Ethics of Artificial Intelligence. UNESCO. https://www.unesco.org/en/articles/recommendation-ethics-artificial-intelligence.

van der Maaten, L. (2022). *Advances in Dimensionality Reduction and Visualization*. Springer, New York.

Vaswani, A., Shazeer, N., Parmar, N., Uszkoreit, J., Jones, L., Gomez, A. N., ... & Polosukhin, I. (2023). Attention is all you need. *Journal of Machine Learning Research*, 21(1), 1–54.

Vedula, V., *et al.* (2023). GPU-accelerated CFD for real-time hemodynamic visualization. *IEEE Transactions on Medical Imaging*, 42(3), 601–615.

Vehlow, C., Reinhardt, T., & Weiskopf, D. (2022). Visual analysis of dynamic networks. In *Visual Informatics* (pp. 117–145). Springer, New York.

Wang, C., Zhang, H., & Li, X. (2022). Immersive visualization in Earth system science education: A review. *Journal of Geoscience Education*, 70(2), 145–159. https://doi.org/10.1080/10899995.2021.2014514.

Wang, C., Zhang, H., & Li, X. (2023). Immersive VR/AR in scientific exploration. *Journal of Visualization*, 26(2), 123–140. https://doi.org/10.1007/s12650-023-00889-1.

Ward, M., Grinstein, G., Keim, D. (2015). *Interactive Data Visualization: Foundations, Techniques, and Applications*. CRC Press. Boca Raton, FL

Ware, C. (2020). *Information Visualization: Perception for Design* (4th ed.). Morgan Kaufmann, San Francisco.

WHO. (2023). Global Health Observatory: Disease burden and visualization tools. World Health Organization. https://www.who.int/data/gho.

Wilkinson, L. (1999). *The Grammar of Graphics* (1st ed.). Springer, New York.

Wilkinson, M., Dumontier, M., Aalbersberg, I. *et al.* (2016). The FAIR Guiding Principles for scientific data management and stewardship. *Sci. Data* 3, 160018. https://doi.org/10.1038/sdata.2016.18.

Wimberly, M. (2023). *Geographic Data Science with R: Visualizing and Analyzing Environmental Change*. Chapman & Hall/CRC Data Science Series. CRC Press, Boca Raton, FL, USA.

WMO. (2021). Guide to meteorological instruments and methods of observation (WMO-No. 8). World Meteorological Organization. https://public.wmo.int/en/standards-and-guidelines/wmo-no-8.

Yeo, B. T., *et al.* (2020). The organization of the human cerebral cortex estimated by intrinsic functional connectivity. *Journal of Neurophysiology*, 123(3), 1188–1209.

Zhang, L., Chen, S., & Li, Q. (2023a). Design principles for public science visualization: A case study of climate change communication. *Journal of Science Communication*, 22(3), A07. https://doi.org/10.22323/2.22030207.

Zhang, Y., Wang, Z., Zhang,J., Shan, G.,Tian, D. (2023b). A survey of immersive visualization: Focus on perception and interaction. *Visual Informatics*, 7(4), 22–35. https://doi.org/10.1016/j.visinf.2023.10.003.

Zhang, Q., *et al.* (2022). Medical visualization for public health communication. *Journal of Visual Communication in Medicine*, 45(2), 103–116.

Zhang, Q., Li, J., Singh, V. P., & Shi, P. (2021). AI-assisted visualization in climate and biomedical sciences. *Environmental Modelling & Software*, 142, 105087. https://doi.org/10.1016/j.envsoft.2021.105087.

Zhang, Q., Li, J., Singh, V. P., & Shi, P. (2021). Development of a web-GIS based platform for real-time flood monitoring and visualization. *Natural Hazards*, 105(1), 1–20. https://doi.org/10.1007/s11069-020-04339-6.

Zhao, Y., Chen, X., & Liu, J. (2023). Multiscale data fusion for geoscience visualization: Methods and applications. *ISPRS Journal of Photogrammetry and Remote Sensing*, 196, 153–168. https://doi.org/10.1016/j.isprsjprs.2023.01.005.

Zhao, Y., Chen, X., & Liu, J. (2023). Universal scientific visualization for public engagement. *Science Communication*, 45(3), 345–367. https://doi.org/10.1177/10755470231156789.

Zheng, Z., *et al.* (2022). Complete reconstruction of the Drosophila brain connectome. *Nature*, 601, 565–574.

Zhou, Y., *et al.* (2023). Interactive rehabilitation visualization improves stroke recovery outcomes. *Journal of NeuroEngineering and Rehabilitation*, 20(1), 15.

Index

G

H

I

O

P

Q

R

S

www.ingramcontent.com/pod-product-compliance
Lightning Source LLC
LaVergne TN
LVHW051052210926
848486LV00002B/89